Math Contests for High School
Volume 6

School Years: 2006-2007 through 2010-2011

Written by

Steven R. Conrad • Daniel Flegler • Adam Raichel

Published by MATH LEAGUE PRESS
Printed in the United States of America

Cover art by Bob DeRosa

First Printing, 2011
Copyright © 2011
by Mathematics Leagues Inc.
All Rights Reserved

Math League Press
P.O. Box 17
Tenafly, NJ 07670-0017

ISBN 978-0-940805-20-0

Preface

Math Contests—High School, Volume 6 is the sixth volume in our series of problem books for high school students. The first five volumes contain contests given in the school years 1977-1978 through 2005-2006. Volume 6 contains contests given from 2006-2007 through 2010-2011. (Use the order form on page 70 to order any of our 18 books.)

These books give classes, clubs, teams, and individuals diversified collections of high school math problems. All of these contests were used in regional interscholastic competition throughout the United States and Canada. Each contest was taken by about 80 000 students. In the contest section, each page contains a complete contest that can be worked during a 30-minute period. The convenient format makes this book easy to use in a class, a math club, or for just plain fun. In addition, detailed solutions for each contest also appear on a single page.

Every contest has questions from different areas of mathematics. The goal is to encourage interest in mathematics through solving *worthwhile* problems. Many students first develop an interest in mathematics through problem-solving activities such as these contests. On each contest, the last two questions are generally more difficult than the first four. The final question on each contest is intended to challenge the very best mathematics students. The problems require no knowledge beyond secondary school mathematics. No knowledge of calculus is required to solve any of these problems. From two to four questions on each contest are accessible to students with only a knowledge of elementary algebra. Starting with the 1992-93 school year, students have been permitted to use any calculator without a QWERTY keyboard on any of our contests.

This book is divided into four sections for ease of use by both students and teachers. The first section of the book contains the contests. Each contest contains six questions that can be worked in a 30-minute period. The second section of the book contains detailed solutions to all the contests. Often, several solutions are given for a problem. Where appropriate, notes about interesting aspects of a problem are mentioned on the solutions page. The third section of the book consists of a listing of the answers to each contest question. The last section of the book contains the difficulty rating percentages for each question. These percentages (based on actual student performance on these contests) determine the relative difficulty of each question.

You may prefer to consult the answer section rather than the solution section when first reviewing a contest. The authors believe that reworking a problem, knowing the answer (but *not* the solution), often helps to better understand problem-solving techniques.

Revisions have been made to the wording of some problems for the sake of clarity and correctness. The authors welcome comments you may have about either the questions or the solutions. Though we believe there are no errors in this book, each of us agrees to blame the others should any errors be found!

Steven R. Conrad, Daniel Flegler, & Adam Raichel, contest authors

Acknowledgments

For the beauty, cleverness, and breadth of his numerous mathematical contributions for the past 30 years, we are indebted to Michael Selby.

For her continued patience and understanding, special thanks to Marina Conrad, whose only mathematical skill, an important one, is the ability to count the ways.

For demonstrating the meaning of selflessness on a daily basis, special thanks to Grace Flegler.

To Daniel Will-Harris, whose skill in graphic design is exceeded only by his skill in writing *really* funny computer books, thanks for help when we needed it most: the year we first began to typeset these contests on a computer.

Table Of Contents

School Year	Contest #	Page # for Problems	Page # for Solutions	Page # for Answers	Page # for Difficulty Ratings
2006-2007	1	2	34	66	68
2006-2007	2	3	35	66	68
2006-2007	3	4	36	66	68
2006-2007	4	5	37	66	68
2006-2007	5	6	38	66	68
2006-2007	6	7	39	66	68
2007-2008	1	8	40	66	68
2007-2008	2	9	41	66	68
2007-2008	3	10	42	66	68
2007-2008	4	11	43	66	68
2007-2008	5	12	44	66	68
2007-2008	6	13	45	66	68
2008-2009	1	14	46	66	68
2008-2009	2	15	47	66	68
2008-2009	3	16	48	66	68
2008-2009	4	17	49	66	68
2008-2009	5	18	50	66	68
2008-2009	6	19	51	66	68
2009-2010	1	20	52	67	68
2009-2010	2	21	53	67	68
2009-2010	3	22	54	67	68
2009-2010	4	23	55	67	68
2009-2010	5	24	56	67	68
2009-2010	6	25	57	67	68
2010-2011	1	26	58	67	68
2010-2011	2	27	59	67	68
2010-2011	3	28	60	67	68
2010-2011	4	29	61	67	68
2010-2011	5	30	62	67	68
2010-2011	6	31	63	67	68

The Contests
. . . .
October, 2006 – March, 2011

HIGH SCHOOL MATHEMATICS CONTESTS

Math League Press, P.O. Box 17, Tenafly, New Jersey 07670-0017

Contest Number 1   *Any calculator without a QWERTY keyboard is allowed.* Answers must be exact *or* have 4 (or more) significant digits, correctly rounded.   **October 24, 2006**

Name _____ Teacher _____ Grade Level _____ Score _____

Time Limit: 30 minutes   *Answer Column*

1-1.	Which positive integer n satisfies $n^{2006} + 2n^{2007} = 3$?	1-1.
1-2.	In non-equilateral isosceles triangle T, the length of each side is an integer. What is the least possible perimeter of T?	1-2.
1-3.	Buses A and B always arrive on time: A every 16 minutes, B every 9 minutes. I take 6 minutes to walk to A's stop and 10 minutes to walk to B's. I leave at random times, and I always walk to the same stop. To which stop (A or B) should I walk to minimize the *expected* total amount of time I'd take to walk to that stop and then wait for a bus to arrive?	1-3.
1-4.	As shown, four cubes (with respective volumes 1, 8, 27, and 125) are attached to each other in a way that minimizes the total surface area of the resulting configuration. What is the total surface area of this configuration?	1-4.
1-5.	The equation $53 = (8 \times 5) + (8 + 5)$ shows how to represent 53 as the product plus the sum of the same two positive integers. What is the least integer greater than 53 which *cannot* be represented this way?	1-5.
1-6.	If $x = 2^{12} \times 3^6$ and $y = 2^8 \times 3^8$, what integer z satisfies $x^x y^y = z^z$? (Note: Your answer should not be expanded. Instead, write your answer as a product of powers of primes.)	1-6.

© 2006 by Mathematics Leagues Inc.

Contest Number 2 *Any calculator without a QWERTY keyboard is allowed.* Answers must be exact *or* have 4 (or more) significant digits, correctly rounded. **November 28, 2006**

Name _____ Teacher _____ Grade Level _____ Score _____

Time Limit: 30 minutes | *Answer Column*

2-1. What is the only number less than $\sqrt{2006}$ whose square is 2006?	2-1.
2-2. A rectangle is divided into two congruent trapezoids, as shown at the right. If both trapezoids have legs of lengths 4 and 5, and both trapezoids have a shorter base of length 3, what is the area of the rectangle?	2-2.
2-3. One 4-tuple that satisfies $a^2+b^2+c^2+d^2 = abcd$ is (2,2,2,2). What is the largest number x for which $(2,2,2,x)$ satisfies $a^2+b^2+c^2+d^2 = abcd$?	2-3.
2-4. I filled 49 packages with big and/or small pens. When 2 packages had the same number of pens, I filled them with different selections of pens. For example, a package could have been filled with 3 pens in only 4 ways: 3 big, 2 big & 1 small, 1 big & 2 small, and 3 small. If no package was left empty, what is the least *total* number of pens I could have put in the 49 packages?	2-4.
2-5. What is the only pair of real numbers (a,b) which satisfies both $$a^3+ab^2 = 30 \text{ and } b^3+ba^2 = 90?$$	2-5.
2-6. What is the area of a semicircular region tangent to two sides of a unit square, with endpoints of its diameter on the other two sides?	2-6.

For 2-2 figure: 3 (top), 4 (left), 5 (middle), 4 (right), 3 (bottom)

Contest Number 3   *Any calculator without a QWERTY keyboard is allowed.* Answers must be exact *or* have 4 (or more) significant digits, correctly rounded.   **January 9, 2007**

Name _____ Teacher _____ Grade Level _____ Score ____

Time Limit: 30 minutes   *Answer Column*

3-1. What is the smallest positive integer c for which $\sqrt{a} + \sqrt{b} = \sqrt{c}$ has a solution in positive integers, no two of which are equal?	3-1.
3-2. For any integer $k > 1$, $k!$ is the product of every integer from 1 to k. Thus, $3! = 1 \times 2 \times 3$. Write the only integer $n > 32$ that satisfies $$(n-23)!(23)! = (n-32)!(32)!$$	3-2.
3-3. For what value of n does $(2^{2007} - 2^{2006})(2^{1997} - 2^{1996}) = 2^n$?	3-3.
3-4. In the accompanying diagram, the lengths of some segments are shown. What is the area of the larger right triangle?	3-4.
3-5. Red birds and blue birds, 30 in all, feasted on worms in my yard. The red birds got 108 worms in total, as did the blue birds. Each blue bird got 3 fewer worms than each red bird. How many blue birds were there?	3-5.
3-6. If r_1, r_2, r_3, r_4 are the roots of $x^4 - 4x^2 + 2 = 0$, what is the value of $$(1+r_1)(1+r_2)(1+r_3)(1+r_4)?$$	3-6.

Contest Number 4 *Any calculator without a QWERTY keyboard is allowed.* Answers must be exact *or* have 4 (or more) significant digits, correctly rounded. **February 6, 2007**

Name _____ Teacher _____ Grade Level ____ Score ____

Time Limit: 30 minutes | *Answer Column*

4-1. I added together some perfect squares, all different. Their total was 55. What was the largest perfect square I added? | 4-1.

4-2. On a trip, nothing exciting was happening, so the four of us each picked a different non-prime positive integer. The greatest common factor of each pair of integers was 1. What is the least possible sum of our four integers? (Note: 1 *is not* a prime.) | 4-2.

4-3. Squares are drawn external to a 3–4–5 triangle, and the new vertices are connected to form hexagon *ABCDEF*, as shown. What is the area of this hexagon? | 4-3.

4-4. Each positive integer can be written as a sum and/or a difference of distinct integral powers of 3 in only one way. For example, $1 = 3^0$, $6 = 3^2 - 3^1$, and $87 = 3^4 + 3^2 - 3^1$. Note that $6 = 3^1 + 3^1$ is *not* valid, since the powers of 3 used are not distinct. Write 2007 as a sum and/or a difference of distinct integral powers of 3. | 4-4.

4-5. A point located on a chord of a circle is 8 cm from one endpoint of the chord and 7 cm from the center of the circle. If a radius of this circle is 13 cm long, how long is the chord, in cm? | 4-5.

4-6. What is the largest possible product of positive integers whose sum is 20? | 4-6.

HIGH SCHOOL MATHEMATICS CONTESTS

Math League Press, P.O. Box 17, Tenafly, New Jersey 07670-0017

Contest Number 5   *Any calculator without a QWERTY keyboard is allowed.* Answers must be exact *or* have 4 (or more) significant digits, correctly rounded.   **March 6, 2007**

Name _____ Teacher _____ Grade Level ____ Score ____

Time Limit: 30 minutes   *Answer Column*

5-1.	It can be proved that there is only one positive integer n for which both n and n^2+2 are prime. What is this unique value of n?	5-1.
5-2.	The squares of the lengths of the sides of a certain right triangle are consecutive integers. What is their sum?	5-2.
5-3.	If a, b, and c are real numbers, what is the only value of a for which the graphs of $y = ax + b$ and $x = ay + c$ are perpendicular to each other?	5-3.
5-4.	The vertices of the shaded triangle are midpoints of three of the six segments into which the medians of a large triangle are divided by their common point of intersection, as shown. If the area of the shaded triangle is 1, what is the area of the large triangle?	5-4.
5-5.	What are both values of x which satisfy $\dfrac{\log_2 x}{\log_4 2x} = \dfrac{\log_8 4x}{\log_{16} 8x}$?	5-5.
5-6.	What are all ordered pairs of integers (a,b), with $0 < a < b$, for which $(\sqrt[a]{2007})(\sqrt[b]{2007}) = \sqrt[9]{2007}$?	5-6.

Contest Number 6   *Any calculator without a QWERTY keyboard is allowed.* Answers must be exact *or* have 4 (or more) significant digits, correctly rounded.   **April 10, 2007**

Name _____ Teacher _____ Grade Level _____ Score _____

Time Limit: 30 minutes   FINAL CONTEST OF THE YEAR   *Answer Column*

6-1. What is the smallest three-digit number that can be written as the product of two consecutive positive integers? | 6-1.

6-2. In my truck, I'm moving a rectangular solid whose volume is 105. If the length of each edge is a prime, what is the total surface area (the sum of the areas of all 6 faces) of the rectangular solid? | 6-2.

6-3. If $x = t^2 - 4t$ and $y = \dfrac{1}{t-2}$, what is the value of x when $y = \dfrac{1}{3\sqrt{2}}$? | 6-3.

6-4. What is the minimum value of $(\sin 2007x)(\cos 2007x)$? | 6-4.

6-5. A function f will be called *repetitive* if there are at least two different values of x in the interval $0 \le x \le 1$ for which $f(x)$ has the same value. What are all real numbers b for which $f(x) = x^2 + bx + 3$ is repetitive? | 6-5.

6-6. Our design team drew all 48 line segments that connect 6 points on one side of rectangle R to 8 points on the opposite side. In at most how many different points inside R can these 48 line segments intersect? | 6-6.

Contest Number 1 *Any calculator without a QWERTY keyboard is allowed.* Answers must be exact *or* have 4 (or more) significant digits, correctly rounded. **October 23, 2007**

Name _____ Teacher _____ Grade Level _____ Score _____

Time Limit: 30 minutes | *Answer Column*

1-1. If $x-y = 2007$ and $y-z = 2008$, what is the value of $x-z$? | 1-1.

1-2. One vertex of a square and the midpoints of the two sides not containing this vertex are vertices of the triangle shaded in the figure at the right. If the area of the square is 16, what is the area of the shaded triangle? | 1-2.

1-3. If $4^x = 10^4$, what is the value of 8^x? | 1-3.

1-4. Andy went bowling with Brandi and Candi. The sum of Andy's score and Candi's score was twice Brandi's score. The sum of Brandi's score and Candi's score was 3 times Andy's score. No one's score was 0. Who had the highest score? | 1-4.

1-5. In the answer space at the right, write the digit 2 four times, positioning the 2s so that the resulting number is as large as possible. (*You may not write any symbol other than a 2.*)

[NOTE: As an example, you could write 222^2; but, by positioning the 2s differently, you can write an even larger number.] | 1-5.

1-6. What are all ordered triples of positive integers (x,y,z) whose product is 4 times their sum, if $x < y < z$? | 1-6.

Contest Number 2 *Any calculator without a QWERTY keyboard is allowed.* Answers must be exact *or* have 4 (or more) significant digits, correctly rounded. **November 20, 2007**

Name _____ Teacher _____ Grade Level _____ Score _____

Time Limit: 30 minutes | *Answer Column*

2-1. If $(x-10)(x+10) = 0$, what is the value of $(x-1)(x+1)$? | 2-1.

2-2. If the least common multiple of the first 2006 positive integers is m, and the least common multiple of the first 2007 positive integers is km, what is the value of k? | 2-2.

2-3. If 2100 words fill any page that a 23-page local newspaper typesets with large type, and 2800 words fill any page that it typesets with small type, how many pages must the paper typeset with small type so that an article of 56 000 words will exactly fill all 23 pages of some future issue? | 2-3.

2-4. The diagonals of convex quadrilateral Q are perpendicular. If three consecutive sides of Q have respective lengths 3, 9, and 19, then how long is the fourth side? | 2-4.

2-5. I wrote a sequence of n integers. In this sequence, the sum of any 3 consecutive terms is positive, while the sum of any 4 consecutive terms is negative. What is the largest possible value of n? | 2-5.

2-6. The degree-measure of each base angle of an isosceles triangle is 40. The bisector of one of the base angles is extended through point P on the leg opposite that base angle so that $PA = PB$, as shown. What is $m\angle A$? | 2-6.

HIGH SCHOOL MATHEMATICS CONTESTS

Math League Press, P.O. Box 17, Tenafly, New Jersey 07670-0017

Contest Number 3   *Any calculator without a QWERTY keyboard is allowed.* Answers must be exact *or* have 4 (or more) significant digits, correctly rounded.   **December 18, 2007**

Name _____ Teacher _____ Grade Level _____ Score _____

Time Limit: 30 minutes   *Answer Column*

3-1. If a 2000×2014 rectangle has the same perimeter as square S, how long is each side of S?

3-1.

3-2. Point O is the center of the circle at the right; and a, b, c, and d are the lengths of the legs of the right triangles shown. If $a^2+b^2+c^2+d^2 = 100$, what is the area of this circle?

3-2.

3-3. When Al, Pat, and Di shared 9 identical slices of pizza equally, Al paid for 5 slices and Pat paid for 4 slices. After Al and Pat split the $9 that Di had paid as her fair share for the pizza she ate, all 3 of them had then paid the same amount. How many dollars did Al get from the money Di paid?

3-3.

3-4. What are both values of x for which $\left(3(3^x)-5(5^x)\right)\left(5(3^x)-3(5^x)\right) = 0$?

3-4.

3-5. The positive integers are placed in groups as follows:

$$(1), (2,3), (4,5,6), (7,8,9,10), (11,12,13,14,15), \ldots$$

and so forth, with n consecutive integers in the nth group. What is the first integer in the 100th group?

3-5.

3-6. There's exactly one real number a for which $ax^2+(a+3)x+(a-3) = 0$ has two positive integer solutions for x. What are these values of x?

3-6.

Solutions on Page 42 • Answers on Page 66

HIGH SCHOOL MATHEMATICS CONTESTS

Math League Press, P.O. Box 17, Tenafly, New Jersey 07670-0017

Contest Number 4 *Any calculator without a QWERTY keyboard is allowed.* Answers must be exact *or* have 4 (or more) significant digits, correctly rounded. **January 15, 2008**

Name _____ Teacher _____ Grade Level ____ Score ____

Time Limit: 30 minutes   *Answer Column*

4-1.   If three different positive integers, a, b, and c, satisfy $a^4 + b^3 = c^2$, what is the least possible value of $a + b + c$?

4-1.

4-2.   At my auction, everyone bid once. A bid \$1. The bids of A and B averaged \$1 more than A's bid. The bids of A, B, and C averaged \$1 more than the average of the bids of A and B. The bids of A, B, C, and D averaged \$1 more than the average of the bids of A, B, and C. Finally, the bids of A, B, C, D, and E averaged \$1 more than the average of the bids of A, B, C, and D. What was E's bid, in dollars?

4-2.

4-3.   What is the smallest integer that can be the perimeter of a square whose diagonal exceeds 29?

4-3.

4-4.   Every leap year is divisible by 4 (but not 100, unless it's a multiple of 400). For example, 2000 was a leap year; 1900 was not. How many leap years are there from 2001 through 4001?

4-4.

4-5.   For what ordered pair of positive integers (b,n), with n as small as possible, is $\log_b n$ the sum of the roots of $9^x - 1006(3^x) + 2008 = 0$?

4-5.

4-6.   A cube has all 8 of its corners cut off, leaving the solid shown. If all 24 vertices of the triangles thus formed are connected to each other by line segments, how many of these segments will pass through the interior of this solid?

4-6.

Solutions on Page 43 • Answers on Page 66

11

HIGH SCHOOL MATHEMATICS CONTESTS

Math League Press, P.O. Box 17, Tenafly, New Jersey 07670-0017

Contest Number 5 *Any calculator without a QWERTY keyboard is allowed.* Answers must be exact *or* have 4 (or more) significant digits, correctly rounded. **February 12, 2008**

Name _____ Teacher _____ Grade Level ____ Score ____

Time Limit: 30 minutes   *Answer Column*

5-1. For which positive integer n does $400^2 \times 400^2 = 16^2 \times n^2$?	5-1.
5-2. In a sequence whose first term is 11, each succeeding term is the sum of the digits of the square of the previous term. For example, the second term is $1+2+1 = 4$ and the third term is $1+6 = 7$. What is the 2008th term of this sequence?	5-2.
5-3. If N is the product of three different primes, then its least possible value is $2 \times 3 \times 5 = 30$. If $N < 100$, what is N's largest possible value?	5-3.
5-4. What is the measure of the angle shown, which is the smallest angle that can be formed by extending two sides of a regular nonagon?	5-4.
5-5. If 26 people (whose first names each start with a different letter) line up at random to climb a ladder, what is the probability that Pat is lucky enough to stand next to Dale?	5-5.
5-6. In a triangle, two sides and the median to the third side have respective lengths 5, 13, and x. Write a list of all possible integral values of x.	5-6.

© 2008 by Mathematics Leagues Inc.

Contest Number 6   *Any calculator without a QWERTY keyboard is allowed.* Answers must be exact *or* have 4 (or more) significant digits, correctly rounded.   **March 18, 2008**

Name _____ Teacher _____ Grade Level _____ Score ____

Time Limit: 30 minutes   *Answer Column*

6-1.	In the accompanying diagram, if the perimeter of equilateral triangle ABC is 9 and the perimeter of equilateral triangle CDE is 3, what is the perimeter of quadrilateral $ADEB$?	6-1.
6-2.	For how many real numbers x does $3^x = 6x - 3$?	6-2.
6-3.	Triangle T's vertices are at $(0,0)$, $(5,0)$, and $(1,2)$. What is the slope of the line through $(0,0)$ that divides T into two triangles of equal area?	6-3.
6-4.	In the sequence below, each angle is in radians. What is the largest number of consecutive terms of this sequence that can be positive? $\cos x, \cos(x+1), \cos(x+2), \cos(x+3), \cos(x+4), \cos(x+5), \cos(x+6)$	6-4.
6-5.	If a, b, and c are positive integers, what is the largest value less than 1 representable by $\frac{1}{a} + \frac{1}{b} + \frac{1}{c}$?	6-5.
6-6.	Bricklayers Pat and Lee alternate turns in a game in which each player removes from 1 to 100 bricks from a common pile that initially has 2008 bricks. The player taking the last brick *wins*. If Pat and Lee both play perfectly, and if Pat goes first, then how many bricks must Pat take on his first turn to guarantee a win?	6-6.

HIGH SCHOOL MATHEMATICS CONTESTS

Math League Press, P.O. Box 17, Tenafly, New Jersey 07670-0017

Contest Number 1 *Any calculator without a QWERTY keyboard is allowed.* Answers must be exact *or* have 4 (or more) significant digits, correctly rounded. **October 21, 2008**

Name _____ Teacher _____ Grade Level ____ Score ____

Time Limit: 30 minutes   *Answer Column*

1-1. If $(x-1)(y-1) = 2008$, what is the value of $(1-x)(1-y)$?

1-1.

1-2. The trapezoid at the right has been split into three right triangles, two of which are congruent. If the trapezoid's bases have lengths 3 and 4, how long is the trapezoid's longer leg, the side labeled x in the diagram?

1-2.

1-3. The expression $(n-2)^2 + 7n$ is divisible by 7 when $n = 2$. What is the largest integer $n < 100$ for which $(n-2)^2 + 7n$ is divisible by 7?

1-3.

1-4. I wake up if and only if both of my alarm clocks ring at the same time. My alarm that's 3 minutes fast first rings when it reads 10:14. It then rings every 9 minutes thereafter. My alarm that's 4 minutes fast first rings when it reads 10:09. It then rings every 7 minutes thereafter. What is the correct time when I wake up?

1-4.

1-5. Write $4x^2-9y^2+4x^3+6x^2y$ as a product of two non-constant polynomials with integral coefficients.

1-5.

1-6. When polygon P, seen at the right, is drawn on a 7×27 grid of unit squares, it passes once through every vertex of every unit square in the grid. What is P's perimeter?

1-6.

© 2008 by Mathematics Leagues Inc.

HIGH SCHOOL MATHEMATICS CONTESTS

Math League Press, P.O. Box 17, Tenafly, New Jersey 07670-0017

Contest Number 2 *Any calculator without a QWERTY keyboard is allowed.* Answers must be exact *or* have 4 (or more) significant digits, correctly rounded. **November 18, 2008**

Name _____ Teacher _____ Grade Level ____ Score ____

Time Limit: 30 minutes | *Answer Column*

2-1. What is the only negative integer x for which $|x+2| = |x+4|$? | 2-1.

2-2. Into how many unit squares can we partition the 36-sided equilateral polygon shown if its perimeter is 36 and if each of its sides is perpendicular to both sides adjacent to it? | 2-2.

2-3. I have 10 nickels, 10 dimes, and 10 quarters. In how many different ways can I pay for a 45¢ item using exact change? (Note: Two different ways use different numbers of nickels, dimes, or quarters.) | 2-3.

2-4. I cut a length of tape into 251 segments, and then I cut each of these 251 segments into 8 smaller pieces. What is the maximum number of cuts that I could make to turn the initial length of tape into the final 2008 pieces? (Note: Each cut completely severs the tape.) | 2-4.

2-5. If exactly two different linear functions, f and g, satisfy $f(f(x)) = g(g(x)) = 4x+3$, what is the product of $f(1)$ and $g(1)$? | 2-5.

2-6. Two triangles which are *not* congruent can actually have five pairs of congruent parts! For example, triangles with side-lengths 8, 12, 18 and 12, 18, 27 have five pairs of congruent parts (two sides and three angles). If two non-congruent *right triangles* have five pairs of congruent parts, what is the ratio of the length of the hypotenuse of either triangle to the length of that triangle's shorter leg? | 2-6.

Solutions on Page 47 • Answers on Page 66

Contest Number 3   *Any calculator without a QWERTY keyboard is allowed.* Answers must be exact *or* have 4 (or more) significant digits, correctly rounded.   **December 16, 2008**

Name _____   **Teacher** _____   **Grade Level** _____   **Score** ____

Time Limit: 30 minutes   *Answer Column*

3-1.   For what value of x will I get 2^{2008} when I double 2^x?

3-1.

3-2.   What per cent discount, applied to a $50 price, yields the same sale price as the total sale price of two items marked at $20 and $30 but discounted 30% and 20% respectively?

3-2.

3-3.   What is the maximum value of $(ab + ad) + (cb + cd)$ if a, b, c, and d have the values 2, 3, 4, and 5, but *not* necessarily in that order?

3-3.

3-4.   What is the perimeter of an equilateral triangle in which the distances from an interior point to the sides are $3\sqrt{3}$, $4\sqrt{3}$, and $5\sqrt{3}$?

(**NOTE**: *A theorem useful in solving this problem is illustrated below.*)

3-4.

3-5.   Line segment ℓ connects the midpoint of the hypotenuse of a 6-8-10 right triangle to a point P on the extension of the triangle's shorter leg, as shown. If the length of ℓ is 8, what is $m\angle P$?

3-5.

3-6.   The coefficients of polynomial P are non-negative integers. If $P(1) = 6$ and $P(5) = 426$, what is the value of $P(3)$?

3-6.

HIGH SCHOOL MATHEMATICS CONTESTS

Math League Press, P.O. Box 17, Tenafly, New Jersey 07670-0017

Contest Number 4 *Any calculator without a QWERTY keyboard is allowed.* Answers must be exact *or* have 4 (or more) significant digits, correctly rounded. **January 13, 2009**

Name _____ Teacher _____ Grade Level ____ Score ____

Time Limit: 30 minutes | *Answer Column*

4-1. What is the smallest integer greater than 2009 which can be the perimeter of a square whose sides have integral lengths?

4-1.

4-2. When the 30th Mersenne prime (a prime of the form $2^p - 1$, where p is prime,) was found in 1983, the *Times* said $p = 131\,049$, the *Globe* claimed $p = 132\,049$, the *News* reported $p = 131\,094$, and the *Post* wrote $p = 132\,094$. Only one of these values is correct. What is the correct value of p? [NOTE: This story may be apocryphal!]

4-2.

4-3. If n is positive, and $n^{10} - n^5 = n^5$, what is the value of n^5?

4-3.

4-4. What are all values of $x > 1$ which satisfy $x^{3\sqrt{x}} = \sqrt{x}^{\,x}$?

4-4.

4-5. In the diagram shown at the right, there's exactly one point on the segment connecting (0,2) to (2,0) for which the angles marked 1 and 2 will be congruent. What are the coordinates of this point?

4-5.

4-6. Starting from opposite ends of a street, each of two messenger boys skated at his top speed towards the other's starting point. From the time they passed each other, one messenger took 1 more minute, and the other took 2 more minutes, to reach their respective destinations. How many minutes did it take the faster messenger to skate the entire distance?

4-6.

Solutions on Page 49 • Answers on Page 66

Contest Number 5 *Any calculator without a QWERTY keyboard is allowed.* Answers must be exact *or* have 4 (or more) significant digits, correctly rounded. **February 24, 2009**

Name _____ Teacher _____ Grade Level ____ Score ____

Time Limit: 30 minutes

Answer Column

5-1. As shown, a square piece of paper has a horizontal crease. If I fold the paper in half along the crease and the new rectangle has a perimeter of 18, what is the perimeter of the square?

5-1.

5-2. Each day last week, every member of the Jogging Team jogged the same integral number of km as every other team member (and the team has more than one member). The total of the distances jogged by the team members last Monday was 287 km. Last Wednesday, that total was 492 km. How many team members are there?

5-2.

5-3. What is the least possible positive difference between two three-digit numbers which together use the digits 4, 5, 6, 7, 8, and 9?

5-3.

5-4. Determine the value of $x > 1$ for which $\log_x(2+x) = \log_x 2 + \log_x x$?

5-4.

5-5. One circle is inscribed in, and a second circle is circumscribed about, a regular polygon of 2009 sides. If the polygon's perimeter is 2009, what is the area of the region between the two circles?

5-5.

5-6. All students at the Academy of Music and Math take both music and math. The probability that a student has an A in math is 1/6. The probability that a student has an A in music is 5/12. The probability that a student with an A in math has an A in music plus the probability that a student with an A in music has an A in math is 7/10. What is the probability that a student has A's in both subjects?

5-6.

HIGH SCHOOL MATHEMATICS CONTESTS

Math League Press, P.O. Box 17, Tenafly, New Jersey 07670-0017

Contest Number 6 *Any calculator without a QWERTY keyboard is allowed.* Answers must be exact *or* have 4 (or more) significant digits, correctly rounded. **March 24, 2009**

Name _____ Teacher _____ Grade Level _____ Score _____

Time Limit: 30 minutes | *Answer Column*

6-1. The length of each side of rectangle R is an integer, and the area of R is 2009. What is the largest possible perimeter of R? | 6-1.

6-2. If I place all three operational symbols $+$, $-$, $\times$ in all possible ways into the blanks of the expression 5___4___6___3, one symbol per blank, each resulting expression will have a unique value. What is the largest of these values? | 6-2.

6-3. How many integers $n > 0$ satisfy the inequality $\frac{3}{65} < \frac{1}{n} < \frac{9}{100}$? | 6-3.

6-4. Suppose that the world's total supply of oil would last 1000 more years if consumed at its current rate. If the current rate of consumption were to double after the first year, and it continued to double each succeeding year, then, to the nearest year, how long would this oil last? | 6-4.

6-5. Two rectangles are positioned in space so that the smaller one is perpendicular to the larger, as shown. A longer side of the smaller rectangle is parallel to and midway between both longer sides of the larger, and is also equidistant from both shorter sides, as shown. The lengths of the sides of the larger rectangle are 1 and x; while those of the smaller are 1 and $x/2$, where $x > 1$. For what value of x will $\triangle ABC$ be equilateral? | 6-5.

6-6. If $3\cos x + 4\cos y = 5$ what is the greatest possible value of $3\sin x + 4\sin y$? | 6-6.

Solutions on Page 51 • Answers on Page 66

19

HIGH SCHOOL MATHEMATICS CONTESTS

Math League Press, P.O. Box 17, Tenafly, New Jersey 07670-0017

Contest Number 1 *Any calculator without a QWERTY keyboard is allowed.* Answers must be exact *or* have 4 (or more) significant digits, correctly rounded. **October 20, 2009**

Name _____ **Teacher** _____ **Grade Level** ____ **Score** ____

Time Limit: 30 minutes   *Answer Column*

1-1. What value of x satisfies $x(x-2009) = x(x+2009)$?	1-1.
1-2. My 5 coupons are worth \$1, \$2, \$5, \$6, and \$10, respectively. What are the only two whole-number dollar amounts from \$1 through \$24 that I cannot pay exactly using one or more of these 5 coupons?	1-2.
1-3. An equilateral triangle and a square share a common side, as shown. In non-equilateral $\triangle ABC$, what is $m\angle ACB$?	1-3.
1-4. Write a triple of positive integers (a,b,c) for which $28a+30b+31c = 365$.	1-4.
1-5. In a certain two-person game, each player, in turn, removes 1, 2, 3, 4, or 5 toothpicks from a common pile, until the pile is exhausted. The player who takes the last toothpick loses. If the starting pile contains 300 toothpicks, how many toothpicks must the first player take on the first turn to guarantee a win with perfect subsequent play?	1-5.
1-6. Last year, each of Big Al's five brothers gave a gift of money to Big Al. The dollar amounts were consecutive integers, and their sum was a perfect cube. If the brothers decide to give Big Al cash gifts with those same properties both this year and next year as well, but this year's sum is larger than last year's, and next year's sum larger still, what is the least possible dollar amount Big Al could get next year from his five brothers combined?	1-6.

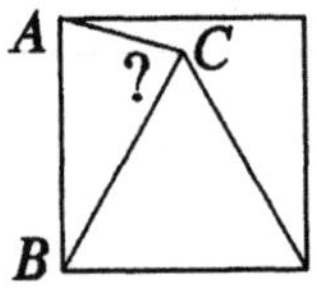

(1-3 figure: square with vertices, A and C at top, B at bottom left, with angle marked "?" at C.)

Contest Number 2 *Any calculator without a QWERTY keyboard is allowed.* Answers must be exact *or* have 4 (or more) significant digits, correctly rounded. **November 17, 2009**

Name _____ Teacher _____ Grade Level ____ Score ____

Time Limit: 30 minutes | *Answer Column*

2-1. My bottle weighs 1450 g when filled with midnight oil. When half empty, my bottle weighs 800 g. How many grams does my bottle weigh when empty?

2-1.

2-2. What are the two integers x for which $x^4 + 4$ is a prime number?

2-2.

2-3. What are all values of x which satisfy $\dfrac{1}{2 + \dfrac{1}{3 + \dfrac{1}{2}}} = \dfrac{1}{2 + \dfrac{1}{3 + \dfrac{1}{x^2 - x}}}$?

2-3.

2-4. In the diagram at the right, a square of area 2009 is divided into a small square and four congruent rectangles. What is the perimeter of one of the four congruent rectangles?

2-4.

2-5. Trapezoid $ABCD$ is isosceles, with $\overline{AD} \parallel \overline{BC}$ and $\overline{AB} \cong \overline{BC} \cong \overline{CD}$. If $\overline{AC} \cong \overline{AD}$, what is $m\angle ADC$?

2-5.

2-6. Al and Bo, who together have $168, bet against each other. Each bets the same fraction of his money as the other bets. If Al wins, he'll have double what Bo then has. If Bo wins, he'll have triple what Al then has. How many dollars does Al have at the start?

2-6.

HIGH SCHOOL MATHEMATICS CONTESTS

Math League Press, P.O. Box 17, Tenafly, New Jersey 07670-0017

Contest Number 3 *Any calculator without a QWERTY keyboard is allowed.* Answers must be exact *or* have 4 (or more) significant digits, correctly rounded. **December 15, 2009**

Name _____ Teacher _____ Grade Level ____ Score ____

Time Limit: 30 minutes | *Answer Column*

3-1. The distance around 3 sides of my rectangular sign is 2009 mm. The distance around a different selection of 3 sides of the same sign is 2011 mm. What is the perimeter of my sign, in mm?

GONE FISHIN'!

3-1.

3-2. As shown, a shaded square has its vertices at points which divide the sides of a large square into segments whose lengths are in a 3:4 ratio. What is the ratio of the area of the shaded square to the area of the large square?

3-2.

3-3. If x is a real number for which $\dfrac{1}{x^3 - 3x^2 + 7x - 5} = \dfrac{5}{6}$, what is the value of $\dfrac{1}{x^3 - 3x^2 + 7x - 4}$?

3-3.

3-4. What rational number x makes $\sqrt{\dfrac{r}{t}\sqrt{\dfrac{t}{r}\sqrt{\dfrac{r}{t}}}} = \left(\dfrac{t}{r}\right)^x$ a true statement for all positive numbers r and t?

3-4.

3-5. The bases of trapezoid T have lengths 10 and x, and the legs have lengths 4 and 5. What are the four positive integers less than 19 which cannot be a value of x?

3-5.

3-6. The 6-digit number $n = ABCDE6$ ends in a 6. Transferring the 6 from last place to first place, and leaving the other digits unchanged relative to one another, results in the same number that one would get by multiplying the original number by 4. Thus, $4 \times ABCDE6 = 6ABCDE$. What is the value of the 6-digit number $n = ABCDE6$?

3-6.

HIGH SCHOOL MATHEMATICS CONTESTS

Math League Press, P.O. Box 17, Tenafly, New Jersey 07670-0017

Contest Number 4 *Any calculator without a QWERTY keyboard is allowed.* Answers must be exact *or* have 4 (or more) significant digits, correctly rounded. **January 12, 2010**

Name _____ Teacher _____ Grade Level ____ Score ____

Time Limit: 30 minutes | *Answer Column*

4-1. If $(10^{10})(201^{10}) = 2010^x$, what is the value of x? | 4-1.

4-2. Alternate vertices of a regular hexagon are connected as shown to form an equilateral triangle. What is the ratio of the sum of the areas of the three shaded regions to the area of the regular hexagon? | 4-2.

4-3. For what value of k will a line whose slope and y-intercept are positive one-digit integers pass through the points (10,73) and (1,k)? | 4-3.

4-4. First, write all the integers from 1 through 30 to form the 51-digit number N = 12345678910111213 14 . . . 282930. Next, remove 44 digits from N without rearranging any of the remaining digits. Of all possible 7-digit results, which is nearest 5 000 000? | 4-4.

4-5. Two knights simultaneously began to ride toward each other. They first met each other after 30 seconds, continued to each other's starting points without meeting again, lost no time in turning, and returned towards each other, each retracing his earlier path. If each knight rode at a steady speed, how many seconds after their first meeting did they meet each other for the second time? | 4-5.

4-6. Equilateral Gothic arch ABC is made by drawing line segment AC, circular arc AB with center C, and circular arc BC with center A. A circle inscribed in this Gothic arch is tangent to $\overset{\frown}{AB}$, $\overset{\frown}{BC}$, and $\overline{AC}$. If $AC = 40$, what is the area of the circle? | 4-6.

Solutions on Page 55 • Answers on Page 67

HIGH SCHOOL MATHEMATICS CONTESTS

Math League Press, P.O. Box 17, Tenafly, New Jersey 07670-0017

Contest Number 5 *Any calculator without a QWERTY keyboard is allowed.* Answers must be exact *or* have 4 (or more) significant digits, correctly rounded. **February 23, 2010**

Name _____ Teacher _____ Grade Level _____ Score _____

Time Limit: 30 minutes   *Answer Column*

5-1.   For what integer n are the roots of $x^2-7x+n = 0$ consecutive integers?

5-1.

5-2.   The first 100 positive integers have a sum of 5050. For how many integers n do the first n positive integers have a sum between 5000 and 6000?

5-2.

5-3.   After I cut the largest circle possible out of a square piece of paper, I discard the leftover scraps of paper. I then cut the largest possible square out of the circular piece of paper and discard the leftover scraps. What fractional part of the original piece of paper remains?

5-3.

5-4.   If $\tan x + \cot x = 4$, what is the value of
$$\sin^2 x + \cos^2 x + \tan^2 x + \cot^2 x + \sec^2 x + \csc^2 x?$$

5-4.

5-5.   Let f and g be functions with respective inverses F and G. An equation relating f and g is $3f(x) = 2g(x)$. If all four functions are defined for all real numbers, and if $F(2010) = G(n)$, what is the value of n?

5-5.

5-6.   As shown, a rectangle and a parallelogram have one vertex in common, and a second vertex of each lies on a side of the other. If each quadrilateral has two sides of length 5, what is the area of the parallelogram?

5-6.

24   Solutions on Page 56 • Answers on Page 67

HIGH SCHOOL MATHEMATICS CONTESTS

Math League Press, P.O. Box 17, Tenafly, New Jersey 07670-0017

Contest Number 6 *Any calculator without a QWERTY keyboard is allowed.* Answers must be exact *or* have 4 (or more) significant digits, correctly rounded. **March 23, 2010**

Name _____ Teacher _____ Grade Level ____ Score ____

Time Limit: 30 minutes   *Answer Column*

6-1. What is the perimeter of the rectangle bounded by the lines $x =$ 2009, $x = -2009$, $y = 2010$, and $y = -2010$?	6-1.
6-2. Starting with 12.5, if we subtract 1, then multiply by 2, we get 23. For what ordered pair of positive one-digit integers (a,b) can we start with 12.5, subtract a, then multiply by b, and get a result of 38?	6-2.
6-3. Watching a baby try to fit a round peg into a square hole and a square peg into a round hole suggested this question: If a circle occupies $A\%$ of its circumscribing square, and a square occupies $B\%$ of its circumscribing circle, which is larger, A or B?	6-3.
6-4. What polynomial is the remainder when $x^{101}+x+1$ is divided by x^2+1?	6-4.
6-5. What is the least possible integer that can be the sum of an infinite geometric progression whose first term is 10?	6-5.
6-6. Semicircles with radii 1 and 2 are externally tangent to each other and internally tangent to a semicircle of radius 3, in the manner shown. How long is a radius of the circle tangent to all three semicircles, as shown?	6-6.

Solutions on Page 57 • Answers on Page 67   25

HIGH SCHOOL MATHEMATICS CONTESTS

Math League Press, P.O. Box 17, Tenafly, New Jersey 07670-0017

Contest Number 1 *Any calculator without a QWERTY keyboard is allowed.* Answers must be exact *or* have 4 (or more) significant digits, correctly rounded. **October 19, 2010**

Name _____ Teacher _____ Grade Level ____ Score ____

Time Limit: 30 minutes | *Answer Column*

1-1. The sum of the three smallest primes and one other prime is 77. What is the product of these four primes? | 1-1.

1-2. What is the least integer greater than 1 million whose square root is also an integer? (Your answer must have exactly 7 digits.) | 1-2.

1-3. What are both integers n for which
$$\left(2^{n^4}\right)\left(2^{n^3}\right)\left(2^{n^2}\right)\left(2^n\right) = 1?$$ | 1-3.

1-4. The cost (in cents) of 8 candies is equal to the number of candies that I can buy for 98 cents. At the same cost per candy, how many cents do 14 candies cost? | 1-4.

1-5. Of the integers between 10^3 and 10^4 that have no repeated digit, how many have digits that increase from left to right? | 1-5.

1-6. In the rectangle at the right, the dotted line segment bisects the obtuse angle through which it is drawn, as shown. What is the length of this dotted line segment, if the other line segments have lengths as marked?

Not drawn to scale

20   x 15

17   8 | 1-6.

Solutions on Page 58 • Answers on Page 67

HIGH SCHOOL MATHEMATICS CONTESTS

Math League Press, P.O. Box 17, Tenafly, New Jersey 07670-0017

Contest Number 2 *Any calculator without a QWERTY keyboard is allowed.* Answers must be exact *or* have 4 (or more) significant digits, correctly rounded. **November 16, 2010**

Name _____ Teacher _____ Grade Level ____ Score ____

Time Limit: 30 minutes

	Answer Column

2-1.   If $x^2 + 6x + 5 = 0$, what is the value of $10x^2 + 60x$?

2-1.

2-2.   For what value of $k > 0$ will the triangle with vertices at $(0,0)$, $(k,0)$, and $(2010,2010)$ have an area of 2010^2?

2-2.

2-3.   In the quadrilateral shown, one of the diagonals is drawn, some of the line segments are marked with their lengths, and some line segments are marked as being perpendicular. What is the perimeter of the quadrilateral?

2-3.

2-4.   A bag of 5 apples, 7 bananas, and 3 carrots costs \$4.41; and a bag of 6 apples, 2 bananas, and 1 carrot costs \$2.37. At these same prices, how much should a bag of 3 apples, 17 bananas, and 7 carrots cost?

2-4.

2-5.   Factor x^4+4; that is, write x^4+4 as a product of two quadratic polynomials with integral coefficients.

2-5.

2-6.   What are all pairs of positive integers (a,b) for which a^2+b exceeds $a+b^2$ by 36?

2-6.

Solutions on Page 59 • Answers on Page 67

27

HIGH SCHOOL MATHEMATICS CONTESTS

Math League Press, P.O. Box 17, Tenafly, New Jersey 07670-0017

Contest Number 3 *Any calculator without a QWERTY keyboard is allowed.* Answers must be exact *or* have 4 (or more) significant digits, correctly rounded. **December 14, 2010**

Name _____ Teacher _____ Grade Level ____ Score ____

Time Limit: 30 minutes   *Answer Column*

3-1. Today, you and I each independently picked a whole number from 1 through 9. Whether we divide my pick by yours or your pick by mine, we get the same remainder. What is this remainder?

3-1.

3-2. What is the least possible perimeter of a triangle whose side-lengths are consecutive perfect squares?

3-2.

3-3. What is the numerical measure of a parallelogram's fourth angle if its other three angles have degree-measures $2x+30$, $3x+50$, and $4x-10$?

3-3.

3-4. What is the smallest positive number y for which $(\sqrt{2})(\sqrt{4})(y)$ and $(\sqrt{3})(\sqrt{6})(y)$ both have positive integral values?

3-4.

3-5. What is the area of the largest circle that can be drawn as shown, tangent to both a quarter-circle (a 90° arc) of length 2π and the line segment that connects the endpoints of that quarter-circle?

3-5.

3-6. At random, I choose 3 different points from among 2010 points evenly spaced on a circle. What is the probability that the 3 points I choose are the vertices of a right triangle?

3-6.

Contest Number 4 *Any calculator without a QWERTY keyboard is allowed.* Answers must be exact *or* have 4 (or more) significant digits, correctly rounded. **January 11, 2011**

Name _____ Teacher _____ Grade Level _____ Score _____

Time Limit: 30 minutes

	Answer Column
4-1. What is the (numerical) area of a square in which two sides have lengths $x-4$ and $x^2-7x+11$?	4-1.
4-2. At least how many dogs do I own if more than half the dogs I own are male, and more than 40% of the dogs I own are female?	4-2.
4-3. From the 99 positive integers less than 100, I chose as many different numbers as I could so that no subset of my numbers had a sum of 100. If the sum of all my numbers was as large as possible, what was the smallest number I actually chose?	4-3.
4-4. A sheet of 8×10 paper is folded along one of its diagonals, with the crease running from corner to corner as shown. What is the length of the dotted line (that part of the longer side hidden from view after the fold)?	4-4.
4-5. For what prime p is $2003p+16$ the square of an integer?	4-5.
4-6. Of all the polynomials f of degree ≥ 1 that have integral coefficients and satisfy $f(\sqrt{3}+\sqrt{2}) = \sqrt{3}-\sqrt{2}$, what is the polynomial of the least degree?	4-6.

HIGH SCHOOL MATHEMATICS CONTESTS

Math League Press, P.O. Box 17, Tenafly, New Jersey 07670-0017

Contest Number 5 *Any calculator without a QWERTY keyboard is allowed.* Answers must be exact *or* have 4 (or more) significant digits, correctly rounded. **February 22, 2011**

Name _____ Teacher _____ Grade Level ____ Score ____

Time Limit: 30 minutes   NEXT CONTEST: MAR. 22, 2011   *Answer Column*

5-1. How long is a side of an equilateral triangle whose area equals the sum of the areas of two equilateral triangles with sides 3 and 4?

5-1.

5-2. If m and n are different positive integers, and if $\dfrac{\frac{1}{m}-\frac{1}{n}}{1-\frac{1}{mn}} = 1$, what is the value of m?

5-2.

5-3. Four girls raced scooters. Alice said, "I was first." Barb said, "I was not last." Cathy said, "I was last." Di said, "I was neither first nor last." If three girls told the truth and one girl lied, who came in first?

5-3.

5-4. A 2011-sided regular polygon has both its inscribed and circumscribed circles drawn. If the length of each side of the polygon is 2, what is the area of the region between the two circles?

5-4.

5-5. In my 6×8 rectangular garden, the paths, shown unshaded, have equal widths. My garden's planting regions are shown as shaded rectangles. If the total areas of the shaded and unshaded regions are equal, how wide is each garden path?

5-5.

5-6. What are all values of k for which at least one pair of real numbers (x,y) satisfies $\sin x + \cos y - \sin x \cos y = k$?

5-6.

© 2011 by Mathematics Leagues Inc.

Contest Number 6   *Any calculator without a QWERTY keyboard is allowed.* Answers must be exact *or* have 4 (or more) significant digits, correctly rounded.   **March 22, 2011**

Name _____ Teacher _____ Grade Level ____ Score ____

Time Limit: 30 minutes   *Answer Column*

6-1. The least common multiple of the first 5 positive integers is 60. What is the least common multiple of the first 6 positive integers?

6-1.

6-2. What is the integer r for which $x^3-49x^2+140x-92 = (x-1)(x-2)(x-r)$ for all values of x?

6-2.

6-3. What is the greatest positive integer that can be the value of x in a triangle whose sides have lengths $\log 4$, $\log 503$, and $\log x$?

6-3.

6-4. The numbers on Al's, Bo's, and Cy's uniforms add up to 21. Cy's number (the largest) exceeds Bo's number by as much as Bo's exceeds Al's. If Cy's were 1 more and Bo's were 1 less, then the ratio of Cy's new number to Bo's new number would be the same as the ratio of Bo's new number to Al's number. What is Al's number?

6-4.

6-5. Although your class has more kids than mine, we can both make the statement below. Altogether how many kids are in our two classes?

In my class, each boy has as many dollars as the number of boys; and each girl has as many dollars as the number of girls. Altogether, the kids in my class have a total of $697.

6-5.

6-6. Connect each of two opposite vertices of a rectangle to the midpoints of both sides not containing that vertex, as shown. If the area of the rectangle is 360, what is the area of the region thus determined (shown shaded)?

6-6.

Complete Solutions

October, 2006 – March, 2011

Problem 1-1

If $n > 1$, the sum on the left > 3. Since n is a positive integer, we see by inspection that $n = \boxed{1}$.

Problem 1-2

In any non-equilateral isosceles triangle with integer-length sides, the legs cannot both have length 1. But if both of T's legs have length 2, the perimeter of T can be as little as $1+2+2 = \boxed{5}$.

Problem 1-3

Since I leave at random times, my wait for bus A is between 0 and 16 minutes, so my waiting time for bus A averages 8 minutes. Similarly, my waiting time for bus B is between 0 and 9 minutes, so it averages 4.5 minutes. The expected amount of time it takes is the sum of the walking time and the waiting time. I expect to take $6+8 = 14$ minutes if I go to bus A and $10+4.5 = 14.5$ minutes if I go to bus B. For the shorter expected time, use bus $\boxed{A}$.

Problem 1-4

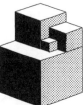

Method I: The edge-lengths of the 4 cubes are 1, 2, 3, 5. Each cube has 6 faces. The areas of the faces are $1^2, 2^2, 3^2, 5^2$. If every face were visible, the surface area would be $6(1+4+9+25) = 234$. The 3 smaller cubes lie on a face of the largest cube, so those 3 faces, with area $1+4+9 = 14$ are hidden. The same amount, 14, is hidden on the top face of the large cube. Similarly, a face of each of the 2 smallest cubes is hidden by a face of the 2nd largest cube, and the smallest cube has a face hidden by a face of the 2nd smallest cube. The total visible surface area of the entire figure is $6(1+4+9+25)-2(1+4+9)-2(1+4)-2(1) = \boxed{194}$.

Method II: The (white) top & (hidden) bottom both have surface area 25. The (dotted) right & (hidden) left both have surface area $25+9 = 34$, and the (black) front & (hidden) back both have surface area $25+9+4 = 38$. The total is $2(25+34+38) = 194$.

Method III: Count every surface of the $5 \times 5 \times 5$ cube, since the top face appears in full (but some of those white parts are raised up, since they are tops of the other cubes). So far, that gives us $6 \times 5^2 = 150$. For the $3 \times 3 \times 3$ cube, add the area of every face except the top and bottom. We already counted the top, the bottom isn't seen, and (for later use) parts of the shaded 3×3 face are seen as sides of the smaller cubes, but we'll count them now. That's another $4 \times 3^2 = 36$.

On the $2 \times 2 \times 2$ cube we already counted the white and shaded sides as parts of the $5 \times 5 \times 5$ and $3 \times 3 \times 3$ cubes, respectively. Now we count its black face and the face opposite that for an extra $4 \times 2 = 8$.

Every visible face of the $1 \times 1 \times 1$ cube has already been counted: the top from the $5 \times 5 \times 5$, the shaded from the $3 \times 3 \times 3$, and the black from the $2 \times 2 \times 2$. Nothing is added. The surface area is $150+36+8 = 194$.

Problem 1-5

Let our integer be $n = ab+a+b = a(b+1)+b$. Although I can't factor $a(b+1)+b$ as it is, I can if I add 1, since $a(b+1)+b+1$ has the factor $(b+1)$. Therefore, $a(b+1)+b+1 = (a+1)(b+1) = n+1$. This tells us something about n: it says $n+1$ *is factorable!* Let's see how to use this. How can I represent 54? First check if $n+1$, 55, is factorable. It is, so we continue. We factor $55 = 5 \times 11$. Let $5 = a+1$, let $11 = b+1$. Then $a = 4$, $b = 10$, and $4 \times 10 + (4+10) = 54$, as required. The first time I run into a problem is when the next integer is a prime, which is not factorable. The first problem occurs when $n = \boxed{58}$.

Problem 1-6

Since $x = 2^{12}3^6$ and $y = 2^8 3^8$, it's easy to see that $x^x y^y = (2^{12}3^6)^x (2^8 3^8)^y = 2^{12x}3^{6x}2^{8y}3^{8y} = 2^{12x+8y}3^{6x+8y}$. Now, $6x = 6(2^{12}3^6) = (2^1 3^1)(2^{12}3^6) = 2^{13}3^7$, and $8y = 8(2^8 3^8) = 2^3(2^8 3^8) = 2^{11}3^8$ and $12x = 2(6x) = 2^{14}3^7$. The factor common to $12x$, $8y$, and $6x$ is $2^{11}3^7$. In fact, $12x = 2^{14}3^7 = 8(2^{11}3^7)$, while $8y = 3(2^{11}3^7)$ and $6x = 2^{13}3^7 = 4(2^{11}3^7)$. Finally, we have $x^x y^y = 2^{12x+8y}3^{6x+8y} = 2^{11(2^{11}3^7)}3^{7(2^{11}3^7)} = (2^{11}3^7)^{(2^{11}3^7)} = z^z$ as long as $z = \boxed{2^{11}3^7}$.

Contests written and compiled by **Steven R. Conrad & Daniel Flegler**   **Mathematics Leagues Inc., © 2006**

Problem 2-1

If $x^2 = \sqrt{2006}$, then $x = \pm\sqrt{2006}$. We are told that $x < \sqrt{2006}$, so $x = \boxed{-\sqrt{2006}}$.

Problem 2-2

As shown in the diagram at the right, the area of the rectangle is $4 \times 9 = \boxed{36}$.

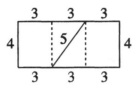

Problem 2-3

$2^2 + 2^2 + 2^2 + d^2 = 8d$, so $d^2 - 8d + 12 = (d-6)(d-2) = 0$. Solving, $d = 2$ or $d = \boxed{6}$.

Problem 2-4

There are 2 ways to package 1 pen (1 big, 1 small), 3 ways to package 2 pens (2 big, 1 of each, 2 small), and 4 ways to package 3 pens (3 big, 2 big & 1 small, 1 big & 2 small, 3 small). In general, there are $n+1$ ways to package n pens. To let me make a total of $2+3+4+5+6+7+8+9 = 44$ packages, I'd need $2(1) + 3(2) + 4(3) + 5(4) + 6(5) + 7(6) + 8(7) + 9(8) = 240$ pens, or more. To make 49 packages, I needed at least $5(9) = 45$ more pens. The total number of pens I used was at least $240 + 45 = \boxed{285}$.

Problem 2-5

Since $b(b^2 + a^2) = 90$ and $a(a^2 + b^2) = 30$, $b/a = 3$, or $b = 3a$. Substituting, $a(a^2 + 9a^2) = 30 \Leftrightarrow 10a^3 = 30 \Leftrightarrow a = \sqrt[3]{3} \Leftrightarrow (a,b) = \boxed{\left(\sqrt[3]{3}, 3\sqrt[3]{3}\right)}$.

Problem 2-6

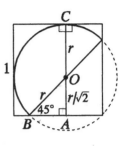

$\triangle BAO$ is an isos. rt. triangle with hypotenuse $OB = r$, so $OA = r/\sqrt{2}$. Since $OC = r$, $AC = 1$, and $AC = AO + OC$, $1 = r + (r/\sqrt{2})$. Hence, $r = 2 - \sqrt{2}$. The area is $\pi r^2/2 = \boxed{\pi(3 - 2\sqrt{2})} \approx 0.5390$.

NOTE: Below are two proofs that $m\angle OBA = 45$. The official solution above doesn't prove this "obvious" fact. It claims, without proof, that $\triangle BAO$ is isosceles. The proofs, though quite elusive, are easy to follow. [Try to write your own proof before reading on.]

Method I: As seen in the diagram, the two small right triangles each have one leg of length $1-r$. Since these right triangles also have $\cong$ hypotenuses, these two triangles are $\cong$, so $AB = A'B'$. Since $PAOA'$ is a square, $AP = A'P$. Thus, $BP = B'P$ and $\triangle BPB'$ is isosceles.

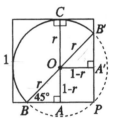

Method II: Call the square L. Draw a radius from the center of the semicircle to each of its two points of tangency on square L. These radii complete a smaller square that we'll call S. Draw the diagonal of L that contains its upper left vertex. Squares S and L, and the quarter-circle contained in square S, are all symmetric across the diagonal of S, so they are also symmetric across the diagonal of L. That means the entire circle and square L are also symmetric across the diagonal, so all corresponding points on the bottom and right sides of square L are reflections of each other across L's diagonal. Therefore, the legs of the right triangle inscribed in the dotted semicircle are congruent, so the right triangle is isosceles.

Contests written and compiled by Steven R. Conrad & Daniel Flegler   © 2006 by Mathematics Leagues Inc.

Problem 3-1

In $\sqrt{a} + \sqrt{b} = \sqrt{c}$, you get the least c when you use the least possible values of a and b. Using $a = 1$ and $b = 4$, $\sqrt{1} + \sqrt{4} = 1+2 = 3 = \sqrt{9}$, so $c = \boxed{9}$.

Problem 3-2

Method I: Since $(n-23)!(23)! = (n-32)!(32)!$, let's equate factorials: $(23)! \neq (32)!$, let's try $(n-23)! = (32)!$, so $n-23 = 32$ and $n = 55$. For this value of n, we also have $(23)! = (n-32)!$, so $n = \boxed{55}$.

Method II: Since $(n-23)!(23)! = (n-32)!(32)!$, we have $\dfrac{n!}{(n-23)!(23)!} = \dfrac{n!}{(n-32)!(32)!} \Leftrightarrow \binom{n}{23} = \binom{n}{32}$. The theorem $\binom{n}{r} = \binom{n}{n-r}$ suggests $n = 23+32 = 55$.

Problem 3-3

Method I:

$$(2^{2007} - 2^{2006}) \times (2^{1997} - 2^{1996})$$
$$= (2^{2006})(2-1) \times (2^{1996})(2-1)$$
$$= (2^{2006}) \times (2^{1996})$$
$$= 2^{2006+1996}$$
$$= 2^{4002}, \text{ so } n = \boxed{4002}.$$

Method II: $2^{2007} = 2(2^{2006}) = 2^{2006} + 2^{2006}$, so $2^{2007} - 2^{2006} = 2^{2006} + 2^{2006} - 2^{2006} = 2^{2006}$. Likewise, $2^{1997} - 2^{1996} = 2^{1996}$. Finally, $2^{2006} \times 2^{1996} = 2^{4002}$.

Problem 3-4

Method I: The area of the smaller rt. $\triangle$ is $\dfrac{3 \times 4}{2} = 6$. If A is the area of the larger rt. $\triangle$, use area ratios of $\sim \triangle$s to get $\dfrac{A}{6} = \left(\dfrac{9}{5}\right)^2$, so $A = \boxed{\dfrac{486}{25}} = \boxed{19.44}$.

Method II: In the diagram at the right, $x^2+y^2 = 34$ and $(x+5)^2+y^2 = 81$. Solving, $x = 2.2$, $y = 5.4$. Now, the triangle's area is $(7.2 \times 5.4)/2 = 19.44$

Problem 3-5

If there were b blue birds and r red birds, we know that $\dfrac{108}{b} + 3 = \dfrac{108}{r}$, or $36r + br = 36b$. Altogether, there are 30 birds, so $r = 30-b$. Substituting, we get $36(30-b) + b(30-b) = 36b$, or $b^2 + 42b - 1080 = (b-18)(b+60) = 0$. Clearly, $b = \boxed{18}$.

Problem 3-6

Method I: Since r_1, r_2, r_3, r_4 are roots of $x^4 - 4x^2 + 2 = 0$, $x^4 - 4x^2 + 2 = f(x) = (x-r_1)(x-r_2)(x-r_3)(x-r_4)$. Notice: $f(-1) = (-1-r_1)(-1-r_2)(-1-r_3)(-1-r_4) = (1+r_1)(1+r_2)(1+r_3)(1+r_4) = (-1)^4 - 4(-1)^2 + 2 = \boxed{-1}$.

Method II: Each $1+r_k$ is a root of $(x-1)^4 - 4(x-1)^2 + 2$. The product $(1+r_1)(1+r_2)(1+r_3)(1+r_4)$ is the product of the roots of this polynomial, which is the constant term $(0-1)^4 - 4(0-1)^2 + 2 = 1-4+2 = -1$.

Method III: If $y = x^2$, then $x^4 - 4x^2 + 2 = 0$ becomes $y^2 - 4y + 2 = 0$, and $y = 2 \pm \sqrt{2}$. Taking the square roots of both sides, $x = \pm\sqrt{2 \pm \sqrt{2}}$. Let $r_1 = \sqrt{2 + \sqrt{2}}$, $r_2 = -\sqrt{2 + \sqrt{2}}$, $r_3 = \sqrt{2 - \sqrt{2}}$, and $r_4 = -\sqrt{2 - \sqrt{2}}$. Then, $(1+r_1)(1+r_2) = 1 + r_1 + r_2 + r_1 r_2 = 1 + r_1 r_2 = -1 - \sqrt{2}$. Similarly, $(1+r_3)(1+r_4) = -1 + \sqrt{2}$. The total product is $(-1-\sqrt{2})(-1+\sqrt{2}) = 1-2 = -1$.

Method IV: Expanding, $(x-r_1)(x-r_2)(x-r_3)(x-r_4) = x^4 - x^3(r_1+r_2+r_3+r_4) + x^2(r_1r_2+r_1r_3+r_1r_4+r_2r_3+r_2r_4+r_3r_4) - x(r_1r_2r_3+r_1r_2r_4+r_1r_3r_4+r_2r_3r_4) - r_1r_2r_3r_4 = x^4 - 4x^2 + 2$.

Now equate coefficients (some 0) of like terms. Then, $(1+r_1)(1+r_2)(1+r_3)(1+r_4) = 1 + (r_1+r_2+r_3+r_4) + (r_1r_2+r_1r_3+r_1r_4+r_2r_3+r_2r_4+r_3r_4) + (r_1r_2r_3+r_1r_2r_4+r_1r_3r_4+r_2r_3r_4) + r_1r_2r_3r_4 = 1 + 0 - 4 + 0 + 2 = -1$.

Method V: Graph the function. Use the graphing calculator to approximate the value of the roots and continue.

Contests written and compiled by Steven R. Conrad & Daniel Flegler   © 2007 by Mathematics Leagues Inc.

Contest # 4 — *Answers & Solutions* — 2/6/07

Problem 4-1

Since $1+4+9+16+25$ and $0+1+4+9+16+25$ are the only ways to get different perfect squares to have a sum of 55, the largest square I added was $\boxed{25}$.

Problem 4-2

The smallest non-prime positive integers are 1 and 4. The next smallest non-prime odd number is 9. The next smallest are 15 and 21, but both have a factor in common with 9, so we'll use 25 instead. Now add. The least possible sum is $1+4+9+25 = \boxed{39}$.

Problem 4-3

Method I: First, surround the hexagon by a 10×11 rectangle. Next, remove the hexagon. You'll be left with 4 "corner" triangles. Their total area is $6+12+6+12 = 36$. The area of the hexagon itself is $10 \times 11 - 36 = \boxed{74}$.

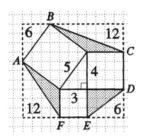

Method II: The 3-4-5 right $\triangle$ has an area of 6. The three squares have respective areas 9, 16, and 25. Connect the squares' vertices as shown to get the 3 shaded $\triangle$s. To show that each $\triangle$ has an area of 6, use a horizontal or vertical side of each $\triangle$ as a base, and drop an altitude from the vertex across from that base. The total area of the hexagon = the sum of all these areas = $6+(9+16+25)+(6+6+6) = 74$.

Problem 4-4

Method I: Begin with a table of powers of 3: $3^0 = 1$; $3^1 = 3$; $3^2 = 9$; $3^3 = 27$; $3^4 = 81$; $3^5 = 243$; $3^6 = 729$; $3^7 = 2187$. Alternately add and subtract (just overshoot the target, if needed) till we zero in on 2007. First, $3^7 = 2187$. Subtract the least power of 3 that overshoots 2007: $3^7-3^5 = 2187-243 = 1944$.

Now add the least power of 3 needed to go to or beyond 2007: $3^7-3^5+3^4 = 2187-243+81 = 2025$. Now subtract $3^3 = 27$ to get 1998, then add $3^2 = 9$ to get 2007. Thus, $2007 = \boxed{3^7-3^5+3^4-3^3+3^2}$.

Method II: In base three, the number is 2202100. Write this out, replacing each "2" by "3−1." Then expand. The result is: $2202100 = 2\times3^6 + 2\times3^5 + 2\times3^3 + 3^2 = (3-1)3^6 + (3-1)3^5 + (3-1)3^3 + 3^2 = 3^7-3^6+3^6-3^5+3^4-3^3+3^2 = 3^7-3^5+3^4-3^3+3^2$.

Problem 4-5

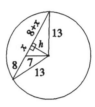 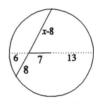 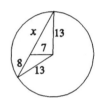

Method I   Method II   Method III

Method I: Using the Pythagorean Thm. twice, we get $h^2 = 7^2-x^2 = 13^2-(8+x)^2$, so $x = 3.5$. Chord $= \boxed{23}$.

Method II: $(6) \times (7+13) = (8) \times (x-8)$, so $x = 23$.

Method III: By the law of cosines, the 7-8-13 $\triangle$'s obtuse angle is 120°. Its supplement, in the 7-13-x $\triangle$, is 60°. Use the law of cosines in the 7-13-x $\triangle$ to get $x = 15$. The chord's length is $8+15 = 23$.

Problem 4-6

We need only integers ≥ 2. We need only integers ≤ 4 (replace integers > 4 with 2's and/or 3's to increase the product). Any 4 can be replaced by two 2's, so we need only 2's and 3's. We need at most two 2s (since $2\times2\times2 < 3\times3$, while $2+2+2 = 3+3$). The maximum product is $3^6\times2 = \boxed{1458}$.

Contests written and compiled by Steven R. Conrad & Daniel Flegler

Problem 5-1

List integer pairs of the form (n, n^2+2) to get: $(1,3)$, $(2,6)$, $(3,11)$, $(4,18)$, $(5,27)$, $(6,38)$, $(7,51)$, ... Since 1 is NOT a prime, the value of n is $\boxed{3}$.

[**NOTE:** We'll prove that $(3,11)$ is the only pair of primes. If $n > 3$ isn't a multiple of 3, then, for some integer $k \geq 1$, either $n = 3k+1$ or $n = 3k+2$. Since both $(3k+1)^2+2 = 9k^2+6k+3$ and $(3k+2)^2+2 = 9k^2+12k+6$ are multiples of 3, neither is a prime. Only for $n = 3$ is n^2+2 a prime.]

Problem 5-2

Method I: The square roots are the lengths of the sides of a right triangle, so the sum of the squares of the shortest lengths must equal the square of the longest length. The squares of the square roots are the consecutive integers themselves, so, for these three consecutive integers, the sum of the two smallest is equal to the largest. The only such triple of consecutive integers is 1, 2, 3. Their sum is $1+2+3 = \boxed{6}$.

Method II: $(\sqrt{x})^2 + (\sqrt{x+1})^2 = (\sqrt{x+2})^2$. Square, simplify, and solve, to get $x = 1$. The sum is 6.

Problem 5-3

If $a \neq 0$, then the slopes, a and $\frac{1}{a}$, cannot have a product of -1, so the lines cannot be perpendicular. If $a = \boxed{0}$, the graphs are horizontal and vertical lines.

Problem 5-4

A line segment connecting the midpoints of two sides of a triangle is parallel to the third side and half as long as the third side. Their corresponding sides are parallel, so the shaded triangle and the original triangle are sim-

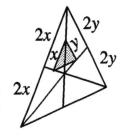

ilar. Since the ratio of their areas is the square of the ratio of the lengths of corresponding sides, it follows that $\frac{a}{A} = \left(\frac{1}{4}\right)^2 = \frac{1}{16}$, so A = $\boxed{16}$.

Problem 5-5

Use the "change of base" theorem: $\log_b a = \dfrac{\log_c a}{\log_c b}$. Three different bases are used. Let's use $c = 2$ as the new base for all the logarithms. The new equation is $\dfrac{(\log_2 4)(\log_2 x)}{\log_2 2 + \log_2 x} = \dfrac{(\log_2 16)(\log_2 4 + \log_2 x)}{(\log_2 8)(\log_2 8 + \log_2 x)}$. Simplifying, $\dfrac{2\log_2 x}{1 + \log_2 x} = \dfrac{4(2 + \log_2 x)}{3(3 + \log_2 x)}$. If we let $t = \log_2 x$, then we get $\dfrac{2t}{1+t} = \dfrac{4(2+t)}{3(3+t)} \Leftrightarrow t^2+3t-4 = (t-1)(t+4) = 0$. If $t = \log_2 x = 1$, then $x = 2$. If $t = \log_2 x = -4$, then $x = 1/16$. The values of x are $\boxed{2, \ 1/16}$.

Problem 5-6

$(2007^{\frac{1}{a}})(2007^{\frac{1}{b}}) = 2007^{\frac{1}{a}+\frac{1}{b}}$, so we must find integer solutions to $\frac{1}{a} + \frac{1}{b} = \frac{1}{9}$, with $0 < a < b$.

Method I: Since $a < b$, it follows that $9 < a < 18$. By trial of the eight possible values of a, we find $a = 10$ or 12, so $(a,b) = \boxed{(10,90), \ (12,36)}$.

Method II: Clearly, $a > 9$ and $b > 9$. Let $a = 9+x$ and let $b = 9+y$, where x and y are positive integers with $0 < x < y$. After substituting, clear the resulting equation of fractions to get $xy = 81$. Finally, $(x,y) = (1,81), (3,27)$, so $(a,b) = (10,90), (12,36)$.

Contests written and compiled by Steven R. Conrad & Daniel Flegler

Problem 6-1

We must use two two-digit numbers, the smallest of which are 10 and 11. Their product is $\boxed{110}$.

Problem 6-2

Since $105 = 3 \times 5 \times 7$, the faces, two of each type, must have dimensions 3×5, 3×7, and 5×7. The total surface area is $2(15+21+35) = \boxed{142}$.

Problem 6-3

Method I: Since $y = \frac{1}{t-2}$ and $y = \frac{1}{3\sqrt{2}}$, it follows that $t-2 = 3\sqrt{2}$. When we square both sides, we get $t^2-4t+4 = (3\sqrt{2})^2 = 18$, so $t^2-4t = 18-4 = \boxed{14}$.

Method II: As above, $t-2 = 3\sqrt{2} \Leftrightarrow t = 3\sqrt{2}+2$. Now, $t^2-4t = t(t-4) = (3\sqrt{2}+2)(3\sqrt{2}-2) = 18-4 = 14$.

Method III: $\frac{1}{y} = t-2$, so $t = \frac{1}{y}+2 = 3\sqrt{2}+2$. Now, $t^2-4t = (3\sqrt{2}+2)^2-4(3\sqrt{2}+2) = 14$.

Problem 6-4

Using the identity $\sin 2\theta = 2\sin\theta\cos\theta$ proves that $\sin 2007x \cos 2007x = \frac{1}{2}(2\sin 2007x \cos 2007x) = \frac{1}{2}(\sin 4014x)$. Half the minimum value of any sine function with an amplitude of 1 is $\boxed{-\frac{1}{2}}$.

Problem 6-5

The graphs of $y = x^2+bx+3$ and $y = x^2$ are translations of each other, so it's easy to see that the graph of $y = x^2+bx+3$ will be *repetitive* when $0 \le x \le 1$ if and only if its minimum occurs when $0 < x < 1$. This parabola's axis of symmetry is $x = -\frac{b}{2}$. Finally, $0 < -\frac{b}{2} < 1$ is equivalent to $\boxed{-2 < b < 0}$.

Problem 6-6

Any 4 points, 2 from one side of the rectangle and 2 from the opposite side, determine a convex quadrilateral whose diagonals cross once inside R. The total number of points of intersection that lie inside R equals $\binom{6}{2} \times \binom{8}{2} = 15 \times 28 = \boxed{420}$.

[**NOTE:** If some of these intersection points are coincident, then there can be fewer than 420 distinct points of intersection.]

Contests written and compiled by Steven R. Conrad & Daniel Flegler　　　©2007 **by Mathematics Leagues Inc.**

Problem 1-1

Since $(x-y)+(y-z) = 2007+2008$, $x-z = \boxed{4015}$.

Problem 1-2

The area of the square is 16, so the length of each side is 4. The unshaded right triangles have areas 4, 4, and 2, so the area of the shaded triangle is $16-10 = \boxed{6}$.

[**NOTE:** In the question, if the word *square* is replaced by the word *parallelogram*, the answer is still 6.]

Problem 1-3

Method I: Take the positive square root of both sides of $4^x = 10^4$ to get $2^x = 10^2$. Now cube both sides to get $8^x = (2^3)^x = (2^x)^3 = (10^2)^3 = \boxed{10^6}$.

Method II: Take the positive square root of both sides of $4^x = 10^4$ to get $2^x = 10^2$. Multiply the two equations to get $8^x = (4^x)(2^x) = (10^4)(10^2) = 10^6$.

Method III: Since $4^{3/2} = 8$, when we raise both sides of $4^x = 10^4$ to the 3/2, we get $(4^x)^{3/2} = (10^4)^{3/2}$, so $8^x = (4^{3/2})^x = (10)^{4(3/2)} = 10^6$.

Method IV: Take the log (base 10) of each side to get $x \log 4 = 4$, or $x = 4/\log 4$. Use your calculator to evaluate $8 \wedge (4 \div (\log 4))$. You'll get $1\,000\,000$.

Problem 1-4

Method I: If we arbitrarily assign A the value 1, then $1+C = 2B$, and $B+C = 3$. Solving, $(A,B,C) = (1, \frac{4}{3}, \frac{5}{3})$; so $A{:}B{:}C = 3{:}4{:}5$. Therefore, $A < B < \boxed{C}$.

Method II: Since $A+C = 2B$ and $B+C = 3A$, subtract the 2nd equation from the 1st to get $A-B = 2B-3A$, so $B = 4A/3 > A$. Substituting this into the 1st equation, $A+C = 8A/3$, so $C = 5A/3 > 4A/3 = B$. Thus, C had the highest score.

Problem 1-5

Write a 2. The next 2 can go either to the right (or to the left) of the first, or it can become an exponent. Since $222^2 > 2222$, we must use exponentiation. If the base is 22, $22^{22} > 22^{2^2} = 22^4 > 222^2$. If the base is 2, then the exponent is 222 or 22^2 or 2^{22} or $2^{2^2} = 2^4$, the largest of which is 2^{22}. Finally, $22^{22} < 32^{22} = (2^5)^{22} = 2^{110} < 2^{2^{22}}$, so the largest of all the possibilities is $\boxed{2^{2^{22}}}$.

[**NOTE:** a^{b^c} is understood to mean $a^{(b^c)}$. See, for example, p.28 of *The Lore of Large Numbers* by P. J. Davis (New Mathematical Library) or p.25 of *Mathematics and the Imagination*. In general, $a^{(b^c)} \neq (a^b)^c$. Some calculators don't handle this correctly!]

Problem 1-6

$\left.\begin{array}{l} xyz = 4(x+y+z) > 4z, \text{ so } xy > 4. \\ xyz = 4(x+y+z) < 12z, \text{ so } xy < 12. \end{array}\right\}$ Thus, $5 \le xy \le 11$.

If $x = 1$, then $5 \le y \le 11$. If $x = 2$, then $3 \le y \le 5$. If $x = 3$, then $y \ge 4$; but this is impossible, since $xy < 12$.

Method I: When $x = 1$, $xyz = 4(x+y+z) \Leftrightarrow yz = 1+y+z$, from which $z = \frac{4(y+1)}{y-4}$. We know $y < z$. The values of $y \in \{5,6,7,8,9,10,11\}$ that work are 5, 6, 8. When $x = 2$, $2yz = 4(2+y+z)$, so $z = \frac{2(y+2)}{y-2}$, so $y = 3$ or 4. Substitute to get z. The 5 solution triples are $\boxed{(1,5,24),\ (1,6,14),\ (1,8,9),\ (2,3,10),\ (2,4,6)}$.

Method II: The two possible x-values, together with the possible xy-values, determine possible (x,y) pairs. Substitute each pair into $xyz = 4(x+y+z)$. See if $z > y$ is also an integer. If $(x,y) = (1,5)$, then $5z = 4(6+z)$, so $z = 24$. Similarly, if $(x,y) = (1,6)$, then $z = 14$; and if $(x,y) = (1,8)$, $z = 9$. No other value of y makes z an integer when $x = 1$. Finally, if $x = 2$ and $(x,y) = (2,3)$, then $6z = 4(5+z)$, so $z = 10$; and if $(x,y) = (2,4)$, then $8z = 4(6+z)$, so $z = 6$. No other value of y makes z an integer when $x = 2$. The 5 solution triples are $(1,5,24)$, $(1,6,14)$, $(1,8,9)$, $(2,3,10)$, $(2,4,6)$.

Contests written and compiled by Steven R. Conrad & Daniel Flegler   Mathematics Leagues Inc., © 2007

Problem 2-1

Since $0 = (x-10)(x+10) = x^2-100$, $x^2 = 100$. The value of $(x-1)(x+1) = x^2-1$ is $100-1 = \boxed{99}$.

Problem 2-2

Notice that $2007 = 3^2 \times 223$ factors into integers that are already factors of the least common multiple of all the positive integers smaller than 2007. Therefore the least common multiple of the first 2006 positive integers $= m =$ the least common multiple of the first 2007 positive integers $= km$, so $k = \boxed{1}$.

Problem 2-3

If x is the number of pages typeset with large type and y is the number of pages typeset with small type, then $2100x+2800y = 56\,000$, so $3x+4y = 80$. We also know that $x+y = 23$. Solving, $x = 12$ and $y = \boxed{11}$.

Problem 2-4

Let's use the Pythagorean Theorem in each of the four right triangles seen in the diagram. We get $3^2+19^2 = a^2+b^2+c^2+d^2$ and $9^2+x^2 = b^2+c^2+a^2+d^2$. Therefore, $3^2+19^2 = 9^2+x^2$, so $x = \boxed{17}$.

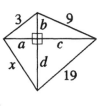

Problem 2-5

Some 5-term sequences are $-2, -3, 6, -2, -3$; or $-2, -2, 5, -2, -2$; or $-3, -3, 7, -3, -3$; or $-3, -3, 8, -3, -3$; or $-7, -4, 12, -6, -5$. Suppose that $c_1, c_2, c_3, c_4, c_5, c_6$ is a 6-term sequence in which the sum of any 3 consecutive terms is positive. Then $c_1+c_2+c_3 > 0$, $c_2+c_3+c_4 > 0$, $c_3+c_4+c_5 > 0$, and $c_4+c_5+c_6 > 0$. Adding these 4 inequalities, and rearranging the terms into groups of 4 consecutive terms, we get $(c_1+c_2+c_3+c_4)+(c_2+c_3+c_4+c_5)+(c_3+c_4+c_5+c_6) > 0$. The inequality in the previous sentence cannot be true if the sum of any 4 consecutive terms is negative, so we conclude that there's no such 6-term sequence. Thus, the largest possible value of n is $\boxed{5}$.

Problem 2-6

In the diagram at the right, reflect B to B', reflecting across the line through A and P as shown. This creates 2 new triangles. $\triangle CPA \cong \triangle CPB'$ by SAS, so $m\angle A = m\angle PB'C = \boxed{80 \text{ or } 80°}$.

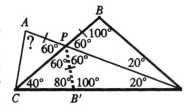

[NOTE: Draw $\overline{AB}$. Its $\perp$ bisector will pass through P. This creates six 60° angles at P and six $\triangle$s (in three pairs of $\cong \triangle$s).]

Contests written and compiled by Steven R. Conrad & Daniel Flegler   ©2007 by Mathematics Leagues Inc.

Problem 3-1

Method I: The rectangle's perimeter $= 2(4014) = 8028 =$ the perimeter of $S = 4s$, so $s = \boxed{2007}$.

Method II: The length of each side of the square is the average of the side-lengths of the rectangle $= \frac{2000 + 2014}{2} = 2007$.

Problem 3-2

$100 = (a^2 + b^2) + (c^2 + d^2) = r^2 + r^2$, so $r^2 = 50$, and area $= \pi r^2 = \boxed{50\pi}$.

Problem 3-3

Each person got 3 slices of pizza. Al paid for 5 slices, but he gave 2 of those slices to Di. Di paid $9 for her 3 slices, so she paid $3 per slice. The number of dollars Al got was $\boxed{6 \text{ or } \$6}$.

Problem 3-4

In turn, set each factor equal to 0. Therefore, either Case 1: $3(3^x) - 5(5^x) = 0$, or Case 2: $5(3^x) - 3(5^x) = 0$. In Case 1, $3^{x+1} = 5^{x+1}$, which is true if and only if each exponent is 0, so $x = -1$. In Case 2, $5(3^x) = 3(5^x)$. Dividing by $15 = 3 \times 5$, $3^{x-1} = 5^{x-1}$, so $x = 1$. The two solutions are $\boxed{1, -1}$.

Problem 3-5

The first group contains 1 integer, the second contains 2 integers, the third contains 3 integers, ..., the 99th contains 99 integers. Altogether, the number of integers used in writing all the numbers in the first 99 groups is $1 + \ldots + 99 = (99/2)(1 + 99) = 4950$. The first integer in the 100th group is $\boxed{4951}$.

Problem 3-6

Method I: Label the roots r and s. Their sum is $r+s$ and their product is rs, so we know that $a(x-r)(x-s) = ax^2 - a(r+s)x + ars = ax^2 + (a+3)x + (a-3)$. Equate coefficients of like powers of x: $-ar - as = a+3$ and $ars = a-3$. Adding, $a(-r-s+rs) = 2$. Since $a \neq 0$, we'll divide through by a to get $-r-s+rs = 2$. Solve by trial and error, or here's a formal approach: *For later, note that $s = 1$ is impossible. To solve for r: $s+2 = rs-r = r(s-1)$. Since $s \neq 1$, divide through by $(s-1)$, so $r = \frac{s+2}{s-1} = 1 + \frac{3}{s-1}$. Since r is a positive integer only when $s-1$ is a positive integral divisor of 3, $s-1 = 1$ or 3, $s = 2$ or 4 (these are the same as the values of r). The only possible positive integral roots x are $\boxed{2, 4}$.

Method II: If $a = 0$, only one value of x works. Divide through by $a \neq 0$: $x^2 + \left(\frac{a+3}{a}\right)x + \left(\frac{a-3}{a}\right) = 0$, or $x^2 + \left(1 + \frac{3}{a}\right)x + \left(1 - \frac{3}{a}\right) = 0^{\dagger}$. Call the roots r and s (where r and s are positive integers). Then, $r+s = -1 - \frac{3}{a}$ and $rs = 1 - \frac{3}{a}$. Subtracting, we get $r+s-rs = -2$. Now continue as in Method I* above.

†**Method III:** In $x^2 + \left(1 + \frac{3}{a}\right)x + \left(1 - \frac{3}{a}\right) = 0$ above, let $t = \frac{3}{a}$. We get $x^2 + (1+t)x + (1-t) = 0$. This has integral roots. Since it's discriminant is a perfect square, $(1+t)^2 - 4(1-t) = t^2 + 6t - 3 = (t+3)^2 - 12$ is a perfect square. This means that t is an integer. Since $(t+3)^2$ is a square and $(t+3)^2 - 12$ is also a square, let's look for a perfect square, $(t+3)^2$, which is 12 more than another square. That must be 16, which is 12 more than 4. If $(t+3)^2 = 16$, $t = 1$ or -7. When $t = -7$ in the equation on the second line, we get $x^2 - 6x + 8 = 0$. This has the required positive integral solutions.

Contests written and compiled by Steven R. Conrad & Daniel Flegler   ©**2007 by Mathematics Leagues Inc.**

Problem 4-1

The smallest conceivable choices for a, b, and c are 1, 2, and 3. Since $1^4 + 2^3 = 3^2$, these choices work, so the least possible sum is $1 + 2 + 3 = \boxed{6}$.

Problem 4-2

Using everyone's name as the dollar value of their bid, $A = 1$. The average of A and B is 2, so $B = 3$. The average of A, B, and C is 3; so $C = 5$. The average of A, B, C, and D is 4; so $D = 7$. Finally, the average of A, B, C, D, and E is 5; so $E = \boxed{9 \text{ or } \$9}$.

Problem 4-3

If the length of a diagonal were 29, then each side-length would be $\frac{29}{\sqrt{2}}$. The perimeter would be $\frac{4 \times 29}{\sqrt{2}} = 82.02\ldots$. Thus, the least possible perimeter is $\boxed{83}$. (When each side-length is $20\frac{3}{4}$, the perimeter is 83.)

Problem 4-4

Theoretically (but not actually since the modern calendar did not appear until fairly recently), the number of leap years from 1 through 2000 would be the same as the number from 2001 through 4001. If every 4th year were a leap year, there'd be $\frac{2000}{4} = 500$ leap years. We must subtract the $\frac{2000}{100} = 20$ century years (years divisible by 100), but add back in the $\frac{2000}{400} = 5$ leap years arising once every 400 years. The total number of leap years is $500 - 20 + 5 = \boxed{485}$.

Problem 4-5

$9^x - 1006(3^x) + 2008 = (3^x - 1004)(3^x - 2) = 0$, so $3^x = 1004$ or $3^x = 2$. Thus, $x = \log_3 1004$ or $x = \log_3 2$. Their sum is $\log_3 2008$, so $(b,n) = \boxed{(3, 2008)}$.

Problem 4-6

Each vertex is connected to all 23 other vertices. If we count how many of these 23 line segments lie on the figure's surface, we can subtract to determine how many lie in the figure's interior. Notice that every vertex lies on exactly 2 of the cube's original faces. Pick any face. Call one of its 8 vertices A. On that face, line segments connect A to each of the other 7 vertices. But vertex A appears on 2 faces, and segments connect A to the 2nd face's other 7 vertices. The total number of segments on the 2 faces is not $7 + 7 = 14$, because we counted the common edge of the 2 faces as 2 segments, 1 in each face. So, there are only 6 new segments on the 2nd face. That's a total of $7 + 6 = 13$ line segments on the surface. Since each vertex is connected to all 23 other vertices, $23 - 13 = 10$ of the line segments drawn from each vertex will pass through the figure's interior. Each of the $24 \times 10 = 240$ segments has 2 endpoints, so each has been counted twice, once for each endpoint. The actual number of segments is $\boxed{120}$.

Contests written and compiled by Steven R. Conrad & Daniel Flegler   © 2008 by Mathematics Leagues Inc.

Problem 5-1

$400^2 \times 400^2 = (4^2 \times 100^2)(4^2 \times 100^2) = 16^2 \times 10000^2$, so $n^2 = 10000^2$ and $n = \boxed{10000}$.

Problem 5-2

The sequence is 11, 4, 7, 13, 16, 13, 16, ... All even-numbered terms from the 4th term on are $\boxed{13}$.

Problem 5-3

Method I: Let p represent a prime larger than either of the other two primes in the product being formed. First try products of the form $2 \times 3 \times p$. The largest such product < 100 is $2 \times 3 \times 13 = 78$. Next try $2 \times 5 \times p$. The largest such product < 100 is $2 \times 5 \times 7 = 70$. Every product of either the form $2 \times 7 \times p$ ($p \geq 11$) or the form $3 \times 5 \times p$ ($p \geq 7$) is larger than 100. Thus, the largest possible value of $N < 100$ is $2 \times 3 \times 13 = \boxed{78}$.

Method II: Begin by factoring integers less than 100: $99 = 3 \times 3 \times 11$, $98 = 2 \times 7 \times 7$, 97 is prime, $96 = 2 \times 2 \times 2 \times 2 \times 2 \times 3$, $95 = 5 \times 19$, $94 = 2 \times 47$, $93 = 3 \times 31$, $92 = 2 \times 2 \times 23$, $91 = 7 \times 13$, $90 = 2 \times 3 \times 3 \times 5$, 89 is prime, $88 = 2 \times 2 \times 2 \times 11$, $87 = 3 \times 29$, $86 = 2 \times 43$, $85 = 5 \times 17$, $84 = 2 \times 2 \times 3 \times 7$, 83 is prime, $82 = 2 \times 41$, $81 = 3 \times 3 \times 3 \times 3$, $80 = 2 \times 2 \times 2 \times 2 \times 5$, 79 is prime, and $78 = 2 \times 3 \times 13$.

Problem 5-4   The answer is $\boxed{20}$ or 20°.

Method I:

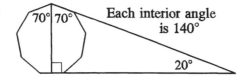

Each interior angle is 140°

Method II: Each interior angle is 140°

Method III:

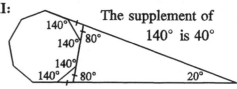

The supplement of 140° is 40°

Method IV:

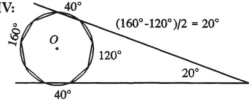

$(160° - 120°)/2 = 20°$

Method V: In the diagram, the pentagon on the right can be split into 3 triangles quite easily, so the sum of the angles of this pentagon is $540° = 40° + 220° + 220° + 40° + 20°$.

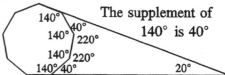

The supplement of 140° is 40°

Problem 5-5

Method I: Line up everyone *except* Pat. There are 24 slots between the 25 people already on line. In addition, Pat can also go to the front or the rear of the line. Of these 26 possible positions for Pat, 2 are next to Dale, so the required probability is $\boxed{\frac{1}{13}}$.

Method II: Treating Pat-Dale as a single entity (with 2 arrangements), the probability is $\frac{2! \, 25!}{26!} = \frac{1}{13}$.

Problem 5-6

Form a parallelogram whose side-lengths are 5 and 13. Any of its half-diagonals is the required median. Let the median's length be x, as shown. Two triangles shown have side-lengths 5, 13, and $2x$. Apply the triangle inequality to either one to get $13 - 5 < 2x < 13 + 5$. Divide by 2 to get $4 < x < 9$. The answers are $\boxed{5, 6, 7, 8}$.

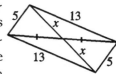

Problem 6-1

The perimeter of quadrilateral $ADEB$ is $3+2+1+2 = \boxed{8}$.

Problem 6-2

The function $f(x) = 3^x$ is increasing and its graph is concave up. The function $g(x) = 6x - 3$ is increasing and its graph is a straight line. The graphs of two such functions can intersect twice, but never more than twice. By observation, these two functions are equal when $x = 1$ or $x = 2$, so the number of solutions is $\boxed{2}$.

Problem 6-3

The line segment through the origin that splits the triangle shown into 2 triangles of equal area is the median drawn from the origin. The side opposite the origin has its midpoint at $(3,1)$, so the slope of the median is $\boxed{\frac{1}{3}}$.

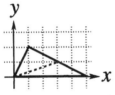

Problem 6-4

Method I: One can see from its graph that the cosine function can't be positive throughout an interval of length $\geq \pi$. In the given sequence, no 5 consecutive terms can all be positive. Here's why: the smallest x–axis interval containing arguments of any 5 terms, such as x, $x+1$, $x+2$, $x+3$, and $x+4$, has length ≥ 4, and $4 > \pi$. When the value of x is slightly greater than $3\pi/2$, such as $x = 4.72$, the first $\boxed{4}$ terms of the original cosine sequence are all positive.

Method II: Use a graphing calculator's table function. Set $y = \cos x$ in radian mode. Use any initial x-value, and use $\triangle x = 1$. A glance at the table shows at most 4 consecutive positive values, so 4 is a good guess!

Problem 6-5

To meet the criteria, clearly $a,b,c \geq 2$. Without loss of generality, we can assume $a \leq b \leq c$. To maximize the sum of their reciprocals (keeping this sum < 1), we want to minimize a,b,c. If $a = 2$, then $\frac{1}{a} = \frac{1}{2}$. Since the sum of all three reciprocals < 1, the sum of the reciprocals of b and c must be less than $\frac{1}{2}$. Thus, both $b,c \geq 3$. If $b = 3$, then $\frac{1}{2} + \frac{1}{3} + \frac{1}{c} < 1$. Therefore, $\frac{1}{c} < \frac{1}{6}$, so set $c = 7$, and $\frac{1}{2} + \frac{1}{3} + \frac{1}{7} = \boxed{\frac{41}{42}}$. Continuing, if $a = 2$ and $b = 4$, then $c > 4$, so use $c = 5$. The sum is $\frac{1}{2} + \frac{1}{4} + \frac{1}{5} = \frac{19}{20} < \frac{41}{42}$. If $b = 5$, then the sum $< \frac{9}{10} < \frac{41}{42}$. If $a = 3$, then the maximum sum is $\frac{1}{3} + \frac{1}{3} + \frac{1}{4} = \frac{11}{12} < \frac{41}{42}$.

Problem 6-6

On the next-to the last round, if Pat leaves 101 bricks, then no matter how many bricks Lee takes (which must be a whole number from 1 to 100 inclusive), Pat can take all the remaining bricks, since the number of bricks remaining will be 100 or less. Similarly, on the round before that, if Pat leaves 202 bricks, then no matter how many bricks Lee takes, Pat can take enough bricks so that 101 bricks remain in the pile at Lee's turn. In general, Pat's winning strategy should be to take enough bricks each round so that the number of bricks remaining is a multiple of 101. If Pat leaves a multiple of 101 bricks each time, then eventually Pat will leave exactly 101 bricks. On Pat's first turn, Pat must remove enough bricks so that the number left is a multiple of 101. Since 2008 is 89 more than a multiple of 101, the number of bricks that Pat should take on the first round is $\boxed{89}$.

[**NOTE:** If the player taking the last brick loses, then, working backwards, Pat must leave 1, 102, 203, . . . , $101n+1$, . . . , 1819, 1920 bricks. Now, Pat's winning strategy is to leave Lee with $101n+1$ bricks. If Lee takes x bricks, $1 \leq x \leq 100$, Pat should take $101 - x$ bricks. Eventually, this leaves Lee with 1 brick. In this case, Pat must first take 88 bricks.]

Contests written and compiled by Steven R. Conrad & Daniel Flegler   ©**2008 by Mathematics Leagues Inc.**

Problem 1-1

Since $(x-1)(y-1) = 2008$, we have $(1-x)(1-y) = (-1)(x-1)(-1)(y-1) = (x-1)(y-1) = \boxed{2008}$.

Problem 1-2

The congruent right triangles have sides of length 3, 4, and 5. In the third right triangle, $x^2 = 5^2 + 5^2$, so $x = \boxed{\sqrt{50}}$.

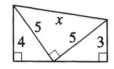

[**NOTE**: The middle triangle can't be $\cong$ to either of the other triangles because its hypotenuse is longer than the hypotenuse of either of those two triangles.]

Problem 1-3

Since n is an integer, and since the 2nd term of the expression $(n-2)^2 + 7n$ is always divisible by 7, the whole expression is divisible by 7 whenever $(n-2)^2$ is a multiple of 7. The square of the integer $(n-2)^2$ is a multiple of 7 if and only if the integer $n-2$ is a multiple of 7. The possibilities are $n-2 = 7, 14, \ldots, 91$, 98. The largest such integer $n < 100$ is $n = \boxed{93}$.

Problem 1-4

Method I: Make a chart that lists the *correct* times when each alarm clock rings. The first clock is 3 minutes fast. When it rings, the correct times are 10:11, 10:20, 10:29, 10:38, **10:47**, ... The second clock is 4 minutes fast. When it rings, the correct times are 10:05, 10:12, 10:19, 10:26, 10:33, 10:40, **10:47**, ... The first common (correct) time is $\boxed{10:47}$.

Method II: Let M be the number of minutes after 10:00 (in the correct time) that the two clocks ring at the same time. Then $M = 11+9x = 5+7y$, or $6 = 7y-9x$, for some non-negative integers x and y. From this last equation, y must be a multiple of 3. The least such y is 6, from which $x = 4$ and $M = 47$.

Problem 1-5

$4x^2 - 9y^2 + 4x^3 + 6x^2y = (4x^2-9y^2) + (4x^3+6x^2y) = (2x-3y)(2x+3y)+(2x^2)(2x+3y) = \boxed{(2x+3y)(2x-3y+2x^2)}$.

Problem 1-6

Let's consider a similar problem applied to a much smaller polygon, say, a 1×1 grid of unit squares, where the polygon is a single unit square. The number of vertices of the unit square is $(1+1)(1+1)$. If there were a 1×2 rectangle, made of 2 unit squares, then the number of vertices would be $(1+1)(2+1) = 2 \times 3 = 6$. Similarly, a 7×27 grid of unit squares will have $(7+1)(27+1) = 8 \times 28 = 224$ vertices. Begin at any of the polygon's vertices and traverse the perimeter of P. You will travel 1 unit each time you go to a new vertex, and the final unit is the return to the first vertex, completing the polygon (which is a closed figure). Hence, the perimeter is $\boxed{224}$.

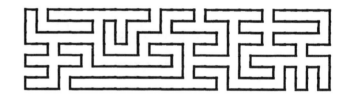

Contests written and compiled by Steven R. Conrad & Daniel Flegler   ©2008 by Mathematics Leagues Inc.

Problem 2-1

Removing the absolute value symbols, either $x+2 = x+4$ (which is impossible) or $x+2 = -(x+4)$. From this equation, $x+2 = -x-4$. Solving, $x = \boxed{-3}$.

Problem 2-2

Method I: To count the boxes quickly, add $2(1+3+5+7) + 9 = 32+9 = \boxed{41}$.

Method II: Since $1+3+5+...+(2n-1) = n^2$, we get $(1+3+5+7) + (1+3+5+7+9) = 4^2+5^2 = 41$.

Problem 2-3

Organize the data. Let's let the triple (n,d,q) represent (# nickels, # dimes, # quarters), with the condition that the total value of the coins is 45¢. Begin by finding all such triples with $q = 0$. The 5 such triples are $(9,0,0)$, $(7,1,0)$, $(5,2,0)$, $(3,3,0)$, and $(1,4,0)$. With $q = 1$, the 3 such triples are $(4,0,1)$, $(2,1,1)$, and $(0,2,1)$. Altogether, there are $\boxed{8}$ such triples.

Problem 2-4

Method I: Each cut through a single thickness of tape creates one additional piece. We began with 1 piece, we ended with 2008 pieces, so the number of cuts we needed to make was at most $\boxed{2007}$.

Method II: To get the 251 pieces requires 250 cuts, at most. To cut each of these pieces into 8 smaller pieces requires 7 cuts, at most. To cut all 251 pieces into 8 smaller pieces each requires 7×251 cuts, at most. We need at most $250 + 7 \times 251 = 2007$ cuts.

Problem 2-5

If $f(x) = ax+b$ is the linear function f, then $f(f(x)) = a(ax+b)+b = 4x+3$, so $a^2x+ab+b = 4x+3$. Equate coefficients of x to get $a^2 = 4$, from which $a = \pm 2$. Equate constant terms to get $ab+b = 3$. Each value of a gives rise to a value of b. Then, $(a,b) = (2,1)$ and $(-2,-3)$, so let $f(x) = 2x+1$ and $g(x) = -2x-3$. Thus, $f(1) = 3$, $g(1) = -5$, and their product is $\boxed{-15}$.

Problem 2-6

The angles are $\cong$*, so the $\triangle$s are similar. Thus, $\frac{a}{b} = \frac{b}{x}$, or $b^2 = ax$. In $\triangle$I, $a^2+b^2 = x^2$, from which $a^2+ax = x^2$. Since

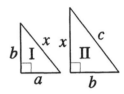

$x > 0$ and $a > 0$, we first use the quadratic formula to solve for x, then we can divide by a. Solving, we get $x = \frac{a + \sqrt{a^2 + 4a^2}}{2}$, so $\frac{x}{a} = \boxed{\frac{1+\sqrt{5}}{2}}$.

*[**NOTE:** Of the 5 pairs of congruent parts, either 3 are angles or 3 are sides. If the 3 congruent pairs were sides, then the triangles would be congruent, and the 3 pairs of angles would also be congruent then. It must be the case that 3 pairs of angles are congruent, but only 2 pairs of sides are congruent.]

[**NOTE:** The answer is known as the *Golden Ratio*.]

[**NOTE:** The hypotenuse of the larger triangle is longer than any side of the smaller triangle, so the legs of the larger triangle are as long as the hypotenuse and longer leg of the smaller triangle. The larger triangle's shorter leg is as long as the smaller triangle's longer leg.]

Contest # 3   *Answers & Solutions*   **12/16/08**

Problem 3-1

Since $2 \times 2^x = 2^{x+1} = 2^{2008}$, $x = \boxed{2007}$.

Problem 3-2

20%($30) + 30%($20) = $6+$6, a $12 total discount on a $50 purchase. As a percent, it's equivalent to a $24 discount on $100. That's $\boxed{24}$ or 24%.

Problem 3-3

Method I: $ab+ad+cb+cd = a(b+d)+c(b+d) = (a+c)(b+d)$. The product is largest when the values of the factors $a+c$ and $b+d$ are equal, so the largest possible value is $(a+c)(b+d) = 7 \times 7 = \boxed{49}$.

[**NOTE**: The *Arithmetic Mean-Geometric Mean Inequality* (for integers) says that if two integers have a fixed sum, their product is maximized when the integers are equal or as close to equal as possible.]

Problem 3-4

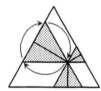 

Here's how to look at the diagrams: The first diagram shows three segments perpendicular to the sides from an interior point. Rotate the encircled shaded triangle 120° so its altitude points to the right, resulting in the second diagram. In the second diagram, the encircled triangle consists of a parallelogram and two shaded triangles. Rotate this encircled triangle 120° to get the third diagram. In the third diagram, the altitudes of all three shaded triangles are vertical. The sum of their lengths is the length of the altitude of the largest triangle shown.

The three diagrams provide a "Proof Without Words" of *Viviani's Theorem: The sum of the distances from any interior point to the sides of an equilateral triangle is the length of the triangle's altitude.* The sum of the lengths of all the altitudes is $12\sqrt{3}$, so the length of one side of the triangle is 24, and its perimeter is $\boxed{72}$.

Problem 3-5

Drop a perpendicular from the midpoint of the hypotenuse of the right triangle to the shorter leg. Since this perpendicular passes through the midpoint of

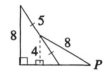

the hypotenuse and is parallel to the longer leg, that perpendicular is half as long as the longer leg. We created a right triangle whose hypotenuse is 8 and whose shorter leg (the perpendicular we dropped) is 4. Thus, $m\angle P = \boxed{30}$ or 30°.

Problem 3-6

If P were a polynomial with non-negative integral coefficients of degree ≥ 4, then $P(5) \geq 5^4 = 625$. Since $P(5) = 426 \neq 625$, we know that P can't be such a polynomial. Let $P(x) = ax^3+bx^2+cx+d$, where some (not all) of the coefficients are 0. Since $P(1) = a+b+c+d = 6$, if any coefficient were 6, the others would be 0. But, when $x = 5$, $P(5) = 6x^n = 6(5^n) = 426$ is impossible when n is an integer. Thus, every coefficient ≤ 5. Can any coefficient be 5? If one coefficient were 5 and another were 1, then the others would be 0. If $a = 5$, then $P(5) > 5x^3 = 625$. This is impossible since $P(5) = 426$. If another coefficient is 5, then $P(x) \leq x^3+5x^2$, so $P(5) \leq 5^3+5(5^2) = 250$. Therefore, no coefficient exceeds 4. Since $426 = 3\times125+2\times25+1$, $(a,b,c,d) = (3,2,0,1)$. This is the only 4-tuple of non-negative integers that satisfies the requirements $P(1) = 6$ and $P(5) = 426$, since base 5 representation is unique if every coefficient ≤ 4; so $P(x) = 3x^3+2x^2+1$, and $P(3) = \boxed{100}$.

Contests written and compiled by Steven R. Conrad & Daniel Flegler   ©2008 by Mathematics Leagues Inc.

Problem 4-1

The smallest integer greater than 2009 that can be the perimeter of a square whose sides have integer lengths is the smallest multiple of 4 larger than 2009. The perimeter will be $\boxed{2012} = 4 \times 503$.

Problem 4-2

Since $1+3+1+0+4+9 = 1+3+1+0+9+4$ are both multiples of 3, the numbers reported by the *Times* and the *News* were divisible by 3. The *Post*'s posting, an even integer, can't be prime, so the correct number, as reported by the *Globe*, was $p = \boxed{132\,049}$.

Problem 4-3

Method I: Since $n > 0$, we can divide each side of $n^{10} = 2n^5$ by n^5 to get $n^5 = \boxed{2}$.

Method II: Since $(n^5)^2 = n^{10}$, we can write the given equation as $(n^5)^2 - n^5 = n^5$, or $(n^5)^2 - 2n^5 = 0$, or $(n^5)(n^5 - 2) = 0$; so $n^5 = 2$.

Problem 4-4

$x^{3\sqrt{x}} = \sqrt{x}^x \Leftrightarrow x^{3\sqrt{x}} = x^{x/2}$. Since $x > 1$, neither base has the value 1, so we can now equate exponents to get $3\sqrt{x} = \frac{x}{2} \Leftrightarrow 9x = \frac{x^2}{4}$. Since $x > 0$, $x = \boxed{36}$.

Problem 4-5

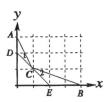

In the diagram seen at the right, $\triangle ACD \sim \triangle BCE$, so $AD{:}BE = 1{:}2 = DC{:}CE$. Thus DC is $\frac{1}{3}$ of DE, so we can conclude that the x-coordinate of C is $\frac{1}{3}$ of the x-coordinate of E. Similarly, the y-coordinate of C is $\frac{2}{3}$ of the y-coordinate of D. The coordinates of point C are $\boxed{\left(\frac{2}{3}, \frac{4}{3}\right)}$.

Problem 4-6

Method I: If t represents the number of minutes each boy skates before passing the other boy, then the slower one took $t+2$ minutes and the faster $t+1$ minutes to travel 1 street-length. Their respective rates, in street-lengths per minute, are $\frac{1}{t+2}$ and $\frac{1}{t+1}$. The total distance traveled by the two boys during the first t minutes is 1 street-length, so $\frac{t}{t+2} + \frac{t}{t+1} = 1$. The positive solution of this equation is $t = \sqrt{2}$, so $t+1 = \boxed{\sqrt{2}+1}$.

Method II: Suppose the faster messenger skates at the rate of r_1 street-lengths per minute and the slower messenger skates at the rate of r_2 street-lengths per minute. The distance each skated before they met equals the distance the other skated after they met. So $tr_1 = 2r_2$ and $tr_2 = 1r_1$, and we have that $2r_2 = tr_1 = t^2 r_2$. Thus, $t = \sqrt{2}$ and the faster messenger took $\sqrt{2}+1$ minutes to skate the entire distance.

Contests written and compiled by Steven R. Conrad & Daniel Flegler   ©2009 by Mathematics Leagues Inc.

Problem 5-1

Method I: If one side of the square has length $2x$, then the square's perimeter is $8x$. When split into two rectangles, each has length $2x$, width x, and perimeter $6x$. The perimeter of each new rectangle is 18, so $x = 3$. The square's perimeter is $8x = 8 \times 3 = \boxed{24}$.

Method II: Trial and error shows that the long side of the rectangle is 6 and the shorter side is 3, so the square's perimeter is $4 \times 6 = 24$.

Problem 5-2

Factoring, $287 = 7 \times 41$ and $492 = 12 \times 41$. The team has more than 1 member; it must have $\boxed{41}$ members.

Problem 5-3

To minimize the difference, d, the hundreds' digits must be consecutive integers. Further, d is minimal when the smaller hundreds' digit is followed by the largest possible tens' digit and the larger hundreds' digit is followed by the smallest possible tens' digit. So far, the integers are 69* and 74*. To minimize d, make the larger remaining digit the units' digit of the smaller number, and make the smaller remaining one the units' digit of the larger number. Finally, $745 - 698 = \boxed{47}$.

Problem 5-4

$\log_x(2+x) = \log_x 2 + \log_x x = \log_x 2x \Leftrightarrow 2+x = 2x$.
Solving, $x = \boxed{2}$.

Problem 5-5

Method I: Let the length of the polygon's apothem (a radius of the inscribed circle) be a. Each side of the polygon has length 1. Use the Pythagorean Theorem to find the length of a radius of the larger circle (the circumscribed circle), as shown. The area of the larger circle is $\pi(a^2 + \frac{1}{4})$ and the area of the smaller circle is πa^2, so the area between the two circles is their difference, $\boxed{\frac{\pi}{4}}$.

Method II: Relabel the lengths in the above diagram so the radii are R and r, with $R > r$. In the right $\triangle$, $R^2 - r^2 = \frac{1}{4}$. To get the difference in the areas of the two circles, multiply this equation through by π.

Problem 5-6

Let the number of students at the Academy be n. Since the probability that a student has an A in math is $\frac{1}{6}$, only $\frac{n}{6}$ students have an A in math. Since the probability that a student has an A in music is $\frac{5}{12}$, only $\frac{5n}{12}$ students have an A in music. The probability that a student with an A in math has an A in music plus the probability that a student with an A in music has an A in math is $\frac{7}{10}$. If x is the number of students who have an A in both, then $\frac{x}{n/6} + \frac{x}{5n/12} = \frac{7}{10}$, so $x = \frac{n}{12}$. Since the number of students who have A's in both is $\frac{n}{12}$, and since the total number of students is n, the probability that a student has A's in both subjects is $\boxed{\frac{1}{12}}$.

Contests written and compiled by Steven R. Conrad & Daniel Flegler   © 2009 by Mathematics Leagues Inc.

Problem 6-1

If a rectangle has a fixed area, its perimeter increases whenever the difference between the lengths of two adjacent sides increases. The perimeter is maximized when one of the rectangle's dimensions is 1 and the other is 2009. That rectangle's perimeter is $\boxed{4020}$.

Problem 6-2

There are six different ways to fill in the blanks in 5___4___6___3 using a different one of the $+$, $-$, and $\times$ symbols in each blank. Those six ways are:

$5 \times 4 + 6 - 3 = 23,$
$5 \times 4 - 6 + 3 = 17,$
$5 + 4 \times 6 - 3 = 26,$
$5 + 4 - 6 \times 3 = -9,$
$5 - 4 \times 6 + 3 = -16,$ and
$5 - 4 + 6 \times 3 = 19.$

Of these six, the largest value is $\boxed{26}$.

Problem 6-3

Method I: Inequalities of positive numbers are reversed when we take reciprocals, as the equivalence of $\frac{1}{4} < \frac{1}{n} < \frac{1}{2}$ and $4 > n > 2$ and $2 < n < 4$ illustrates. Hence, the inequalities $\frac{3}{65} < \frac{1}{n} < \frac{9}{100}$ and $\frac{100}{9} < n < \frac{65}{3}$ are equivalent. Exactly $\boxed{10}$ integers n satisfy $11.\overline{1} < n < 21.\overline{6}$.

Method II: Use a calculator. The given information is approximately $0.046 < \frac{1}{n} < 0.09$. The easiest way to see which integers n satisfy this inequality is to use the calculator's table feature, with $n = x$. Using $y = \frac{1}{x}$ and ΔTbl $= 1$, y is between 0.046 and 0.09 for every integer x from 12 through 21 inclusive.

Problem 6-4

If oil usage doubles each year, then in $\boxed{10}$ years, we'd use $1+2+4+8+16+32+64+128+256+512 = 1023$ years of oil (at this year's usage levels).

Problem 6-5

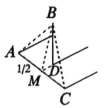

In right $\triangle BDM$, $BD = \frac{x}{2}$, and MD is half the difference of the lengths of the two rectangles, so $MD = \frac{x-1}{2}$ and $MB^2 = \left(\frac{x-1}{2}\right)^2 + \left(\frac{x}{2}\right)^2 = \frac{2x^2 - 2x + 1}{4}$. Each side of $\triangle ABC$ will have a length of 1 only if $MB^2 = \left(\frac{\sqrt{3}}{2}\right)^2 = \frac{3}{4} = \frac{2x^2 - 2x + 1}{4} \Leftrightarrow 2x^2 - 2x - 2 = 0$. Finally, since x is positive, $x = \boxed{\frac{1+\sqrt{5}}{2}}$.

[**NOTE:** The answer is known as the *Golden Ratio*.]

Problem 6-6

Let $3\sin x + 4\sin y = A$. Since $3\cos x + 4\cos y = 5$, square both equations and add results: $A^2 + 25 = 9\sin^2 x + 24\sin x \sin y + 16\sin^2 y + 9\cos^2 x + 24\cos x \cos y + 16\cos^2 y = 25 + 24(\cos x \cos y + \sin x \sin y)$, so $A^2 = 24\cos(x-y)$. At most, $A^2 = 24$ and $A = \boxed{\sqrt{24}}$.

[**NOTE:** If $x = y$, $3\cos x + 4\cos y = 3\cos x + 4\cos x = 7\cos x = 5$, so $x = y = \pm\text{Arc}\cos\frac{5}{7}$. Choosing $x > 0$, $\cos x = \frac{5}{7}$, $\sin x = \frac{\sqrt{24}}{7}$, & $3\sin x + 4\sin y = 7\sin x = \sqrt{24}$. This function's actual and theoretical maximum is $\sqrt{24}$.]

Contests written and compiled by Steven R. Conrad & Daniel Flegler   **©2009 by Mathematics Leagues Inc.**

Problem 1-1

If $x = 0$, both sides are equal by inspection. If $x \neq 0$, then we can divide both sides of the given equation by x to get $x - 2009 = x + 2009$, an equation with no solutions. The only solution is $\boxed{0}$.

Problem 1-2

The sum of the values of all 5 coupons is $24. Taken together, the coupons worth less than $5 are worth only $3, so there's no way to get $4 worth of coupons. We're told that there are only two unachievable amounts, and we know $4 in unachievable. Therefore, the other unachievable amount must be $20, since there's no way to take away $4 from the $24 total to get $20 remaining. Hence, the two unachievable numbers of dollars are $\boxed{4, 20}$, or $4, $20.

Problem 1-3

Since $\angle ABC$ is a right angle minus an angle of the equilateral triangle, $m\angle ABC = 90 - 60 = 30$. The equilateral triangle and the square share a side, so $BA = BC$, and $m\angle ACB = (180 - 30) \div 2 = \boxed{75}$ or 75°.

Problem 1-4

In a non-leap year, every month has 28, 30, or 31 days, and the year itself has 365 days. If you remember this, then it's easy to go through the months of the calendar and find the solution $(a,b,c) = $ **(1,4,7)**. If we replace three 30s with one 28 and two 31s, we'll get the solution **(2,1,9)**. (Replacing one 28 and two 31s with three 30s produces only triples containing a negative integer or a 0.) *Either of the triples shown in bold type above is a correct answer.*

[**NOTE:** Alternatively, notice that c must be an odd number less than 12. Continue with trial and error.]

Problem 1-5

Work backwards. You want to leave your opponent with 1 toothpick and avoid being left with 1 yourself. You can't leave your opponent with 2, 3, 4, 5, or 6 toothpicks or he will be able to remove the right number to leave you with 1 toothpick. It would be great to leave him with 7, because whatever he does from there, you are safe and you will be able to leave him with 1. Now treat leaving your opponent with 7 as the goal; the same logic says that leaving him with 8, 9, 10, 11, or 12 would be bad, but 13 is a good number to leave at the end of your turn. The goal is to leave your opponent with $6n + 1$ toothpicks (since that eventually leaves him with $6(0) + 1 = 1$ toothpick). Once you've left him with $6n + 1$ toothpicks, do this: whenever your opponent takes x toothpicks, you take $6 - x$ toothpicks. The largest integer less than 300 that fits the goal is $6(49) + 1 = 295$, so you should remove $300 - 295 = \boxed{5}$ toothpicks at your first turn.

Problem 1-6

Since the sum of these 5 consecutive positive integers is a perfect cube, $(x-2) + (x-1) + x + (x+1) + (x+2) = 5x$ is a perfect cube. This is possible only if x is of the form $25n^3$, where n is a positive integer. The smallest possible value of n is 1. Last year, the least amount Al could have possibly received was $5 \times \$25 = \$125$. This year, the least amount of money he could receive as a gift is $5 \times \$(25 \times 2^3) = \$1000$. Next year, the least dollar amount that Al can possibly receive is $5 \times (25 \times 3^3) = 3^3 \times 5^3 = 15^3 = \boxed{3375}$ or $3375.

Contests written and compiled by Steven R. Conrad & Daniel Flegler   **© 2009 by Mathematics Leagues Inc.**

Problem 2-1

Method I: When half the midnight oil was removed, the weight loss was 1450 g − 800 g = 650 g. Therefore, the midnight oil weighed 2×650 g = 1300 g. In grams, the weight of the bottle is $\boxed{150}$.

Method II: The bottle plus half the midnight oil weighs 800 g. Doubling, two bottles plus all the midnight oil would weigh 1600 g. Since one bottle plus all the midnight oil weighs 1450 g, the weight of the bottle, in grams, is 1600 − 1450 = 150.

Problem 2-2

By observation, $x = \boxed{\pm 1}$.

[**NOTE:** Let's prove that $x = \pm 1$ are the only integers for which $x^4 + 4$ is a prime. We can factor $x^4 + 4$ by writing $x^4 + 4 = (x^4 + 4x^2 + 4) - (4x^2) = (x^4 + 4x^2 + 4) - (2x)^2 = (x^2 + 2)^2 - (2x)^2 = (x^2 + 2x + 2)(x^2 - 2x + 2)$. The product is a prime, so either $x^2 + 2x + 2 = 1$, and $x = -1$, or $x^2 - 2x + 2 = 1$, and $x = 1$.

Problem 2-3

The two fractions cannot be equal unless $x^2 - x = 2$. Hence, $x^2 - x - 2 = (x+1)(x-2) = 0$, so $x = \boxed{-1, 2}$.

Problem 2-4

Method I: If the small square has side s, and if a short side of each rectangle is r, then a long side of each rectangle is $r + s$, and each rectangle's semiperimeter is $2r + s$.

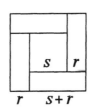

The big square's side is $\sqrt{2009} = 2r + s$, so each rectangle's perimeter is $\boxed{2\sqrt{2009}}$.

Method II: Let ℓ be the length and w the width of a small rectangle. A side of the big square is $\ell + w$, so $(\ell + w)^2 = 2009$, $\ell + w = \sqrt{2009}$, and $2(\ell + w) = 2\sqrt{2009}$.

Problem 2-5

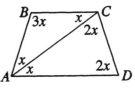

Let $x = m\angle BAC = m\angle BCA$. Alternate-interior angles of parallel lines are congruent, so $m\angle CAD = x$ too. Now, $m\angle CDA = m\angle BAD = 2x$. Finally $m\angle B = m\angle BCD = 3x$. Then, $5x = 180$, so $x = 36$ and $m\angle ADC = 2x = \boxed{72}$ or 72°.

Problem 2-6

Method I: Let Al have $\$a$ and let Bo have $\$b$. Let k be the fraction of his money that each bets, so the actual bets are $\$ka$ and $\$kb$. Since $a + kb = 2(b - kb)$, we know that $b = \frac{a}{2 - 3k}$. Also, $b + ka = 3(a - ka)$, so $a = \frac{b}{3 - 4k}$. Thus, $b = \frac{a}{2 - 3k} = \frac{b}{3 - 4k} \div (2 - 3k)$, so $(2 - 3k) \times (3 - 4k) = 1$. Therefore, $k = 1$ (impossible) or $k = \frac{5}{12}$. Finally, $168 - a = \frac{a}{2 - 3k}$, so $a = \boxed{72}$.

Method II: If Al wins, he'll have twice as much as Bo. Thus, $a + kb = (2/3)(168) = 112$ and $b - kb = 168 - 112 = 56$. If Bo wins, he'll have 3 times as much as Al, so $b + ka = (3/4)(168) = 126$ and $a - ka = 168 - 126 = 42$. Therefore, $238 = 112 + 126 = a + kb + b + ka = a + b + k(a + b) = (a + b)(1 + k)$. We were told that $a + b = 168$, so $1 + k = \frac{238}{168} = \frac{17}{12}$ and $k = \frac{5}{12}$. Now, $a - ka = a(1 - k) = 42$, so $\frac{7}{12}a = 42$, and $a = 72$.

Method III: $3kb = 2b - a$ and $4ka = 3a - b$. Multiplying the 1st equation by $4a$, $12kab = 8ab - 4a^2$. Multiplying the 2nd equation by $3b$, $12kab = 9ab - 3b^2$. Subtract, rearrange, and factor to get $4a^2 + ab - 3b^2 = (4a - 3b) \times (a + b) = 0$. Thus, $a = \frac{3}{4}b$. Since $a + b = 168$, $a = 72$.

Method IV: After deducting their respective bets of $\$a$ and $\$b$, let Al have $\$A$ left and let Bo have $\$B$ left. Since each bet the same fraction of his money, $A/B = a/b$. If the pot contained $\$P$, then $A + P = 2B$ and $B + P = 3A$. Eliminate P to get $2B - A = 3A - B$. Then, $A/B = 3/4 = a/b$, so $a = (3/7)(168) = 72$.

Contests written and compiled by Steven R. Conrad & Daniel Flegler   ©2009 by Mathematics Leagues Inc.

Problem 3-1

Method I: If the rectangle's sides are ℓ and w, with $\ell > w$, then $(2\ell+w)-(\ell+2w) = 2011-2009 = 2$; so $\ell-w = 2$; and $\ell = w+2$. Since $2009 = 2w+\ell = 3w+2$, we get $w = 669$, $\ell = 671$; and the perimeter is $2(\ell+w) = 2(671+669) = \boxed{2680}$ or 2680 mm.

Method II: Add $2\ell+w = 2011$ and $2w+\ell = 2009$ to get $3\ell+3w = 4020$, from which $2(\ell+w) = 2680$.

Problem 3-2

In the diagram at the right, if a side of each small square is x, then a side of the large square is $7x$ and, by the Pythagorean Theorem, a side of the shaded square is $5x$. The ratio of the areas of the shaded square and the largest square is the square of the ratio of their side-lengths. That ratio is $25x^2/49x^2 = 25:49 = \boxed{\frac{25}{49}}$.

Problem 3-3

Method I: Taking reciprocals, $x^3-3x^2+7x-5 = \frac{6}{5}$. Adding 1 to each side, $x^3-3x^2+7x-4 = \frac{11}{5}$. Taking reciprocals one more time, $\dfrac{1}{x^3-3x^2+7x-4} = \boxed{\frac{5}{11}}$.

Method II: Use a graphing calculator to solve the equation for a real solution x_1. Next replace x with x_1 in the fraction whose value is sought. There are shortcuts for carrying out this procedure. Direct substitution is the most laborious method. Graphing calculators feature several function evaluation tools. Learn how to use at least one to make the task easy!

Problem 3-4

Since $\sqrt{\frac{r}{t}\sqrt{\frac{t}{r}\sqrt{\frac{r}{t}}}} = \left(\frac{r}{t}\left(\frac{t}{r}\left(\frac{r}{t}\right)^{\frac{1}{2}}\right)^{\frac{1}{2}}\right)^{\frac{1}{2}} = \left(\frac{r}{t}\right)^{\frac{1}{2}}\left(\frac{t}{r}\right)^{\frac{1}{4}}\left(\frac{r}{t}\right)^{\frac{1}{8}}$, and since $\left(\frac{t}{r}\right)^{\frac{1}{4}} = \left(\frac{r}{t}\right)^{-\frac{1}{4}}$, the product equals $\left(\frac{r}{t}\right)^{\frac{1}{2}}\left(\frac{r}{t}\right)^{-\frac{1}{4}}\left(\frac{r}{t}\right)^{\frac{1}{8}} = \left(\frac{r}{t}\right)^{\frac{3}{8}} = \left(\frac{t}{r}\right)^{-\frac{3}{8}}$; so $x = \boxed{-\frac{3}{8}}$.

[**NOTE:** There's one approach that avoids fractional exponents except at the end. Square both sides of the given equation, and simplify the result. Do this twice more and you'll get $1 = (t/r)^{8x+3}$, so $x = -3/8$.]

Problem 3-5

 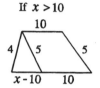

First, $x \neq 10$. (If $x = 10$, then the quadrilateral would be a parallelogram—but a parallelogram cannot have opposite sides with lengths 4 and 5.) The diagrams shown above represent the two possible situations. The resulting inequalities and their solutions are:

$1 < 10-x < 9$	$1 < x-10 < 9$
$-9 < x-10 < -1$	$11 < x < 19$
$1 < x < 9$	

The only positive integers < 19 which are not among these solutions are $\boxed{1, 9, 10, 11}$.

Problem 3-6

Method I: We know that $4(ABCDE6) = 6ABCDE$, so $4(10ABCDE+6) = 600\,000+ABCDE$, from which $39\,ABCDE = 599\,976$. Dividing by 39, $ABCDE = 15384$, so $ABCDE6 = \boxed{153\,846}$.

[**NOTE:** Method II shows that the problem may be solved without knowing how many digits n has.]

Method II: The numbers are of the form $4(\ldots 6) = 6\ldots$. Start by multiplying: $4(\ldots 6) = \ldots 4$, with a carry of 2. Put the 4 to the left of the 6 to get a new trial multiplier, $(\ldots 46)$, which will be multiplied by 4. To multiply $4(\ldots 46)$, we continue by multiplying 4×4, then adding 2 (the carry from 4×6) to get 18. Put the 8 in front to get $(\ldots 846)$, and the carry is 1. Continue this process until there's no carry. In this problem, there's no carry at $4 \times 153\,846 = 615\,384$.

Contests written and compiled by Steven R. Conrad & Daniel Flegler   © 2009 by Mathematics Leagues Inc.

Problem 4-1

If m and n are positive, then $m^x n^x = (mn)^x$; so the product is 2010^{10}, and $x = \boxed{10}$.

Problem 4-2

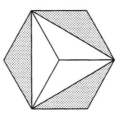

In the diagram, the center of the regular hexagon is connected to the vertices of the equilateral triangle. Of the six congruent triangles in the resulting picture, 3 are shaded, so the shaded region has $\boxed{\frac{1}{2}}$ the hexagon's area.

Problem 4-3

Let the line be $y = mx + b$. Since $(10,73)$ is on the line, substitute to get $73 = 10m + b$. Since m and b are both positive one-digit numbers, $m = 7$, $b = 3$, and $y = 7x + 3$. This line contains $(1,k)$, so $k = 7 + 3 = \boxed{10}$.

Problem 4-4

Let's find the largest number < 5 million. We get
1234567891011121314151617181920212223242526272829 30,
so the largest such number is 4 998 930.

Let's find the smallest number > 5 million. We get
1234567891011121314151617181920212223242526272829 30,
so the smallest such number is 5 001 220.

Since there are not enough 0s to form 5 million, the choice above that is closer to 5 million is $\boxed{4\,998\,930}$.

Problem 4-5

When the knights first pass each other, the total of the distances they've traveled equals the distance that had been between them at the start. By the time they pass each other a second time, the total of all the distances they've traveled then equals three times the distance that had been between them at the start. In other words, the total distance covered to get from the first point at which they pass each other to the second is twice the sum of the distances traveled by the two knights to get from their respective starting points to the point at which they passed each other the first time. It took 30 seconds for them to pass each other the first time, so the number of seconds between the first and second times they passed each other is $\boxed{60}$ or 60 seconds.

Problem 4-6

In the diagram, let a radius of the inscribed circle be r. $\overline{AC}$, the radius of both circular arcs, is 40, so the longer leg of each right triangle is 20, and the hypotenuse is $40-r$. Use the Pythagorean Theorem in either right triangle to get $r^2 + 20^2 = (40 - r)^2$. Solving, we get $r = 15$; so the area of the inscribed circle is $\boxed{225}$.

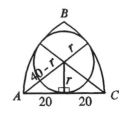

Contests written and compiled by Steven R. Conrad & Daniel Flegler   ©2010 **by MATHEMATICS LEAGUES INC.**

Problem 5-1

Method I: $(x-3)(x-4) = x^2-7x+12$; so $n = \boxed{12}$.

Method II: Let x, $x+1$ be the consecutive solutions. Then $x^2-7x+n = 0$ and $(x+1)^2-7(x+1)+n = 0$ as well. Subtract the 1st equation from the 2nd to get $2x+1-7 = 0 \Leftrightarrow x = 3$, $x+1 = 4$, so $n = 3 \times 4 = 12$.

Problem 5-2

$1 + 2 + \ldots + 99 + 100 = 5050$ and $1 + 2 + \ldots + 108 + 109 = 5995$. As long as $n = 100, 101, 102, 103, 104, 105, 106, 107, 108$, or 109, the sum of the first n positive integers exceeds 5000 and is less than 6000. There are $\boxed{10}$ such values of n.

Problem 5-3

Method I: Draw the inner square's diagonals to get 4 congruent isosceles right triangles in the inner square and 8 in the outer square. Thus, the area of the inner square is $\boxed{\text{half}}$ the area of the outer square.

[**NOTE:** Compare the solutions of #5-3 and #4-2.]

Method II: Let's find the ratio of the areas of the 2 squares in the diagram. Their side-lengths are in the ratio $\sqrt{2}/2$. The ratio of their areas = (the ratio of their side-lengths)$^2 = (\sqrt{2}/2)^2 = \frac{1}{2}$ or 0.5 or 50% or "half."

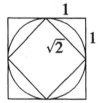

Problem 5-4

The value of $\sin^2 x + \cos^2 x$ is always 1.

Since $\tan^2 x + \cot^2 x = (\tan x + \cot x)^2 - 2$, it follows that $\tan^2 x + \cot^2 x = 4^2-2 = 16-2 = 14$.

Since $\sec^2 x + \csc^2 x = (1 + \tan^2 x) + (1 + \cot^2 x)$, it follows that $\sec^2 x + \csc^2 x = 1 + 1 + 14 = 16$.

The sum of all three is $1 + 14 + 16 = \boxed{31}$.

Problem 5-5

Method I: Since f and F are inverses, $F(2010) = t$ if and only if $f(t) = 2010$. Since $3f(t) = 2g(t)$, it follows that $g(t) = (3/2)f(t)$, so $g(t) = 3015$. Now apply G to both sides to get $t = G(3015)$, so $n = \boxed{3015}$.

Method II: Let $k = 3f(x) = 2g(x)$. Then $f(x) = k/3$. Applying the inverse F to each side, $x = F(k/3)$. Similarly, $g(x) = k/2$, so $x = G(k/2)$. Equating results, $F(k/3) = G(k/2)$. Thus, $F(2010) = F(6030/3) = G(6030/2) = G(3015)$, from which $n = 3015$.

Problem 5-6

A side of the shaded triangle is a side of the rectangle, and one altitude of the shaded triangle is an altitude of the rectangle, so the area of the 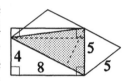 shaded triangle is half the area of the rectangle. In the same way, one side of the shaded triangle is a side of the parallelogram, and the shaded triangle and the parallelogram have the same altitude, so the shaded triangle's area is half the parallelogram's area. Therefore, the area of the parallelogram must equal the area of the rectangle = $5 \times 8 = \boxed{40}$.

[**NOTE 1:** As long as the rectangle's and parallelogram's shorter sides are the same length, then a rotation of the parallelogram can be made around the vertex angle of the isosceles triangle so that its side that's marked with a 5 coincides with the rectangle's side that's marked with a 5. This rotation through an angle equal to the measure of the isosceles triangle's vertex angle will then place the parallelogram's other side of length 5 on the very line that contains the rectangle's left side.]

[**NOTE 2:** Proving that the measure of the isosceles triangle's vertex angle is exactly $2 \arctan \frac{1}{2}$ is a problem in elementary geometry, even though the language used is trigonometry. Just show that one angle is twice another. Now **that's** a nice problem!]

Contests written and compiled by Steven R. Conrad & Daniel Flegler   ©2010 by Mathematics Leagues Inc.

Problem 6-1

The rectangle's width is the distance from $x = -2009$ to $x = 2009$. That width is $2009 - (-2009) = 4018$. Similarly, the height is $2010 - (-2010) = 4020$. The rectangle's perimeter is $2(4018 + 4020) = \boxed{16076}$.

Problem 6-2

Method I: It's best to work backwards and use inverse operations. Begin with 38. We get an odd integer if we divide by 2. We get 9.5 if we divide by 4. Add 3 to get 12.5—our starting number! To verify this, reverse the steps and use inverse operations: subtract 3 to get 9.5, and multiply by 4 to get 38. The answer is $\boxed{(3,4)}$.

Method II: $b(12.5 - a) = 38$ if and only if $b(25 - 2a) = 76$. Since $25 - 2a$ is odd, $b = 4$, $25 - 2a = 19$, and $a = 3$.

Problem 6-3

A circle of radius 1 inscribed in a square of side 2 occupies $(\pi r^2)/s^2 = \pi/4 \approx 78.5\%$ of the square. That percentage is the same for every circle inscribed in a square. Similarly, a square of side $\sqrt{2}$ is inscribed in a circle of radius 1, so every square occupies $2/\pi \approx 63.7\%$ of its circumscribing circle. The answer is $\boxed{A}$.

Problem 6-4

Method I: Do the long division. Watch for a pattern:

$$x^2+1 \overline{)\,x^{101}+0x^{99} + \ldots + 0x^2 + x + 1} \quad \begin{array}{c} x^{99}-x^{97}+x^{95}-x^{93} + \ldots - x + \frac{2x+1}{x^2+1} \end{array}$$

$$\underline{x^{101} + x^{99}}$$
$$-x^{99} + \ldots + 0x^2 + x+1$$
$$\underline{-x^{99}-x^{97}}$$
$$x^{97}+\ldots+0x^2+x+1$$
$$\vdots$$
$$\vdots$$
$$-x^3+x+1$$
$$\underline{-x^3-x}$$
$$\boxed{2x+1} \quad \text{(the remainder)}$$

Method II: When we divide by x^2+1, let's call the quotient and remainder polynomials $Q(x)$ and $R(x)$, respectively. From the division algorithm, we know that $(x^2+1)Q(x) + R(x) = x^{101} + x + 1$. At $x = i$, we get $R(i) = 2i + 1$. The remainder polynomial is of a lower degree than the divisor, so the remainder is $ax+b$, where a and b are real. Hence, $R(x) = 2x+1$.

Method III: Let $f(x) = x^{101}+x+1 = x^{101}-x+2x+1 = x(x^{100}-1) + (2x+1)$. Clearly, $x^4-1 = (x^2+1)(x^2-1)$ is divisible by x^2+1. Since t^n-1 is divisible by $t-1$ for all integers $n > 0$, $x^{4n}-1$ is divisible by x^4-1, which is divisible by x^2+1. Therefore, $x^{4n}-1$ is divisible by x^2+1 for all integers $n > 0$, and thus $x^{100}-1$ is divisible by x^2+1. The first term of $x(x^{100}-1) + (2x+1)$ is divisible by x^2+1, so the remainder will be the right side, $2x+1$.

Problem 6-5

Call the progression's sum S and call its common ratio r, where $|r| < 1$. Since $a_1 = 10$, we have $S = \frac{10}{1-r}$. Since $|r| < 1, S > \frac{10}{1-(-1)} = 5$. The least integer > 5 is $\boxed{6}$.

Problem 6-6

Since the radius of the large circle is 3, the cevian through the interior of the large triangle is $3-r$. In the triangle with angle θ, $(1+r)^2 = 2^2+(3-r)^2 -$ 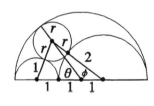 $2(2)(3-r)\cos\theta$ follows from the law of cosines. In the triangle with angle ϕ, we get $(2+r)^2 = 1^2+(3-r)^2 - 2(1)(3-r)\cos\phi$. Double both sides of this equation, then use the fact that, since angles θ and ϕ are supplementary, $\cos\phi = -\cos\theta$. We then get $2(2+r)^2 = 2+2(3-r)^2+2(2)(1)(3-r)\cos\theta$. Add this to the equation derived from the first triangle to get $2(2+r)^2+(1+r)^2 = 6+3(3-r)^2$. Solving, $r = \boxed{\frac{6}{7}}$.

Contests written and compiled by Steven R. Conrad & Daniel Flegler   ©2010 by Mathematics Leagues Inc.

Problem 1-1

Three of the primes are the smallest possible primes. Since 1 is not a prime, three of the four primes must be 2, 3, and 5. The fourth prime is $77 - 5 - 3 - 2 = 67$. The product of all four is $2 \times 3 \times 5 \times 67 = \boxed{2010}$.

Problem 1-2

The square root of 1 million is 1 thousand. Add 1 and square it to get $\boxed{(1001)^2 \text{ or } 1\,002\,001}$.

Problem 1-3

If $(2^{n^4})(2^{n^3})(2^{n^2})(2^n) = 1 = 2^0$, the sum of the exponents is $n^4 + n^3 + n^2 + n = n(n^3 + n^2 + n + 1) = n(n+1)(n^2+1) = 0$. Since n is real, $n = \boxed{0, -1}$.

Problem 1-4

Method I: Let c be the cost of one candy, in cents. Then, $8c = \frac{98}{c}$, so $c^2 = \frac{98}{8} = \frac{49}{4}$. Thus, $c = \frac{7}{2}$, and $14c = \boxed{49 \text{ or } 49\cent}$.

Method II: If n candies can be bought for 98¢, then the number of cents that each candy costs is $\frac{98}{n}$. But n is also the number of cents that it costs to buy 8 candies, so the cost of each, in cents, is $\frac{n}{8}$. Equating, $\frac{98}{n} = \frac{n}{8}$, so $n = \sqrt{2 \cdot 4 \cdot 2 \cdot 49} = 2 \cdot 2 \cdot 7 = 28$. Since 28 is the number of candies that can be bought for 98¢, the cost of 14 candies would be 49¢.

Problem 1-5

For each selection of 4 different digits, there's only one way to arrange them in increasing order, left to right. So the question becomes: how many ways can we select the 4 digits? Since the leftmost digit is a 1 or a larger digit, the chosen digits cannot include a 0. Therefore, we must choose 4 of the 9 digits other than 0. Hence, the total number of integers between 1 thousand and 10 thousand whose digits increase from left to right is $\binom{9}{4} = \boxed{126}$.

Problem 1-6

In the scale diagrams, $\triangle$I and $\triangle$II share a common side that bisects the obtuse angle through which the side is drawn. That makes $\triangle$I $\cong \triangle$II by SAS, so the two dotted lines we added to the diagram are congruent. Next, by the Pythagorean Theorem, $20^2 + y^2 = (25 - y)^2 + 5^2$, from which $y = 5$. In the third diagram, we can use the Pythagorean Theorem to get $x^2 = 12^2 + 20^2 = 544$, so $x = \boxed{\sqrt{544}} = 4\sqrt{34}$.

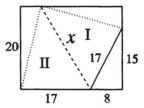

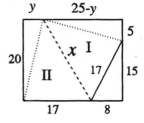

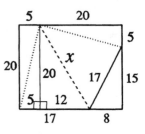

Contests written and compiled by Steven R. Conrad & Daniel Flegler

Problem 2-1

Since $x^2 + 6x = -5$, $10x^2 + 60x = -5 \times 10 = \boxed{-50}$.

Problem 2-2

A quick sketch of the triangle on a pair of rectangular coordinate axes would look like the sketch at the right (which is not drawn to scale). Draw an altitude from (2010,2010) to the x-axis. The length of that altitude is 2010. Thus, the area of the triangle = $2010k/2 = 2010^2$. That will happen when $k = \boxed{4020}$.

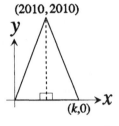

[**NOTE**: If $k < 0$ were permitted, then another solution would be $k = -4020$.]

Problem 2-3

By the Pythagorean Theorem, $(x-9)^2 + (x+8)^2 = (x-7+x)^2$, so $x = 16$ and we get the next diagram.

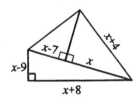 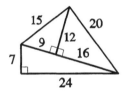

The perimeter of the quadrilateral is $7 + 15 + 20 + 24 = \boxed{66}$.

Problem 2-4

Let's see how we might get the required 3 apples and other fruit from pre-filled bags of 5 apples and other fruit or 6 apples and other fruit. One way is to buy 3 five-apple bags and sell 2 six-apple bags to my friend. In fact, $3(5a+7b+3c = \$4.41) - 2(6a+2b+c = \$2.37)$ gives us the result $3a+17b+7c = \boxed{\$8.49}$.

[**NOTE**: This problem would have no solution unless the contents and price of the new collection of fruit was dependent on the two conditions in the first sentence of the problem. The dependency can be seen in the linear combination $x(5a+7b+3c = \$4.41) + y(6a+2b+c = \$2.37)$, from which $a(5x+6y)+b(7x+2y)+c(3x+y) = \$4.41x+\$2.37y$. We want the value of $3a+17b+7c$, so we try $5x+6y = 3$, $7x+2y = 17$, and $3x+y = 7$. This system is consistent when $x = 3$ and $y = -2$. Finally $\$4.41x + \$2.37y = \$13.23 - \$4.74 = \$8.49$.]

Problem 2-5

$$
\begin{aligned}
x^4 + 4 &= (x^4 + 4x^2 + 4) - (4x^2) \\
&= (x^2 + 2)^2 - (2x)^2 \\
&= (x^2 + 2 + 2x)(x^2 + 2 - 2x) \\
&= \boxed{(x^2 + 2x + 2)(x^2 - 2x + 2)}
\end{aligned}
$$

[**NOTE**: Alternatively, find the fourth roots of -4, then multiply pairs of linear factors with conjugate roots.]

Problem 2-6

$$
\begin{aligned}
(a^2+b) - (a+b^2) &= 36 \Leftrightarrow \\
(a^2-b^2) - (a-b) &= 36 \Leftrightarrow \\
(a+b)(a-b) - (a-b) &= 36 \Leftrightarrow \\
(a-b)(a+b-1) &= 36.
\end{aligned}
$$

Since a and b are positive integers, the larger factor, $(a+b-1)$, is a positive integer. Thus, the factor $(a-b)$ is also positive. One of these factors must be even; the other must be odd. Factoring numerically, we get:

$a + b - 1 =$	36	12	9	(one factor's values)
$a - b =$	1	3	4	(the other factor's values)

So, $2a-1 = $ 37   15   13,

$\qquad a = $ 19   8   7

Finally, $(a,b) = \boxed{(19,18),\ (8,5),\ (7,3)}$.

Contests written and compiled by **Steven R. Conrad & Daniel Flegler**   ©2010 by **Mathematics Leagues Inc.**

Problem 3-1

If we didn't pick the same numbers, then one of us chose a number larger than the other's. When we divide the smaller by the larger, the remainder is the smaller. But if you divide the larger by the smaller, the remainder would have to be smaller than the smaller. OOPS. What happened? Well, we began by supposing we didn't pick the same numbers. That supposition must have been false, so we must have picked the same numbers. The remainder is $\boxed{0}$.

Problem 3-2

We need three consecutive squares in which the largest is less than the sum of the other two. Here's a list: 1   4   9   16   25   36   49. The smallest three that could be sides of a triangle are 16, 25, and 36. The perimeter of that triangle is $\boxed{77}$.

Problem 3-3

Two of the angles must be congruent since the figure is a parallelogram. If $2x+30 = 3x+50$, then $x = -20$, which yields impossible angles. If $2x+30 = 4x-10$, then $x = 20$. If $3x+50 = 4x-10$, then $x = 60$, which is not possible, because the resulting angles are not possible. The three given angles have measures 70, 110, and 70. The degree-measure of the fourth angle is $\boxed{110}$.

Problem 3-4

Let $2y\sqrt{2} = a$ and let $3y\sqrt{2} = b$, where a and b are positive integers. Clearly, $3a = 2b$, so the least possible values of a and b are 2 and 3 respectively. Substitute these into the equations in the first sentence to get $y = \boxed{1/\sqrt{2}}$.

Problem 3-5

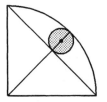

The largest such circle is the circle (of radius r) shown shaded. The quarter-circle (radius R) has arc-length 2π, so $\frac{2\pi R}{4} = 2\pi$ and $R = 4$. By the Pythagorean Thm, the segment connecting the endpoints of the quarter-circle has length $4\sqrt{2}$. In the big right triangle, the median to the hypotenuse is an altitude, so its length is half the hypotenuse $= 2\sqrt{2}$. The diameter of the shaded circle has length $4-2\sqrt{2}$, so $r = (4-2\sqrt{2})/2$. Finally, the area of the shaded circle is $\pi r^2 = \boxed{\pi(6-4\sqrt{2})\text{ or }1.07802416891\ldots}$.

Problem 3-6

Method I: The points will be vertices of a right triangle if and only if two of them are diametrically opposite. There are 2 cases; we'll add their probabilities together. Pick any point. Of the remaining 2009 points, one is diametrically opposite the first. If it is chosen, then the third point may be any other point. That has probability $1 \times \frac{1}{2009} \times 1 = \frac{1}{2009}$. If the second point is not diametrially opposite the first point ($p = 1 - \frac{1}{2009} = \frac{2008}{2009}$), then the third point could be diametrically opposite the first ($p = \frac{1}{2008}$) or second ($p = \frac{1}{2008}$), with probability $1 \times \left(\frac{2008}{2009}\right) \times \left(\frac{1}{2008} + \frac{1}{2008}\right) = \frac{2}{2009}$. Now, add $\frac{1}{2009}$ from before. The sum is our answer, $\boxed{\frac{3}{2009}}$.

Method II: First find the probability that the 3 points do **not** form a rt. $\triangle$. The 1st point can be any point. The 2nd can be any of the 2008 points *not* on the same diameter. The 3rd can be any of the 2006 points *not* on the same diameter as a point already chosen. Thus, prob (rt $\triangle$) $= 1-(1 \times \frac{2008}{2009} \times \frac{2006}{2008}) = \frac{3}{2009}$.

Method III: There are 1005 pairs of diametrically opposite points. We want 1 such pair and another point. The number of ways to do this is 1005×2008. The total number of ways one can select 3 points from among 2010 is $\binom{2010}{3}$. The quotient is $\frac{3}{2009}$.

Contests written and compiled by Steven R. Conrad & Daniel Flegler   ©2010 by Mathematics Leagues Inc.

Problem 4-1

The sides of a square are congruent, so $x^2 - 7x + 11 = x - 4 \Leftrightarrow x^2 - 8x + 15 = (x-3)(x-5) = 0$. When $x = 3$, the sides have negative lengths; so $x = 5$, the length of each side is 1, and the square's area is $\boxed{1}$.

Problem 4-2

The fraction of female dogs is greater than 40% and less than 50%, so its value is between 0.4 and 0.5. The denominator is the number of dogs, so we want it to be as small a whole number as possible. The numerator is the number of female dogs. Trying fractions with a value less than one-half, no number of halves, thirds, fourths, fifths, or sixths has a value less than 0.5 but larger than 0.4. The first success is the fraction $3/7 \approx 0.4286$, so I own $\boxed{7}$ dogs.

Problem 4-3

I chose $\{50, 51, 52, \ldots, 97, 98, 99\}$. Here's why: Among the whole numbers from 1 through 99, there are 49 pairs whose sum is 100 (such as 1 and 99, 2 and 98, $\ldots$, 49 and 51), so I chose at most one number from each such pair. For a maximum sum, I chose the larger number in each pair. Those are the largest 49 numbers. I also chose 50, since it had no pairmate with whom its sum was 100. I couldn't introduce another number without forming a pair whose sum was 100, so my selection's smallest number was $\boxed{50}$.

Problem 4-4

The shorter sides of each of the two small congruent right triangles are congruent. Each is labeled with an x. By the Pythagorean Theorem, $8^2 + x^2 = (10-x)^2$. Solving, $x = 1.8$. Therefore, the length of the dotted line is $10 - x = \boxed{8.2}$.

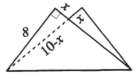

Problem 4-5

The number 2003 is a prime, so if $2003p + 16 = x^2$, we can write $2003p = (x+4)(x-4)$. If $x+4 = 2003$, then $p = 1995$, which is not prime. If $x-4 = 2003$, then $p = x+4 = \boxed{2011}$.

[**NOTE:** None of $x+4 = \pm 1$, $x-4 = -2003$, or $x-4 = \pm 1$ yields a viable solution, but $x+4 = -2003$ also gives us $p = 2011$.]

Problem 4-6

Since $(\sqrt{3} + \sqrt{2})(\sqrt{3} - \sqrt{2}) = 1$, the factors $\sqrt{3} + \sqrt{2}$ and $\sqrt{3} - \sqrt{2}$ are reciprocals. If $r = \sqrt{3} + \sqrt{2}$, we will then have $r^2 = 5 + 2\sqrt{6}$, or $r^2 - 5 = 2\sqrt{6}$. Squaring, $r^4 - 10r^2 + 25 = 24$, or $r^4 - 10r^2 + 1 = 0$. Term by term, if we divide $-r^4 + 10r^2 = 1$ by r, we get the $\frac{1}{r}$ reciprocal term we want. Since $-r^3 + 10r = \frac{1}{r}$, we can say that $f(x) = \boxed{-x^3 + 10x}$.

[**NOTE:** It can be established, by consideration of each case separately, that no linear and no quadratic polynomial works. It can also be proven that the cubic polynomial we found above is unique.]

Problem 5-1

An equilateral $\triangle$ of side-length s has area $\frac{s^2\sqrt{3}}{4} = ks^2$. If two triangles have sides 3 and 4, then the sum of their areas is $k(3^2+4^2) = k(5^2)$. That's the area of an equilateral triangle of side-length $\boxed{5}$.

Problem 5-2

Set the numerator equal to the denominator to get $\frac{1}{m} - \frac{1}{n} = 1 - \frac{1}{mn}$. Multiplying by mn, $n-m = mn-1$. Rearranging, $(n+1) = mn+m = m(n+1)$. Thus, $(n+1) - m(n+1) = (n+1)(1-m) = 0$. Either $n = -1$ (not a positive integer) or $m = \boxed{1}$.

Problem 5-3

Let's examine what each girl said to see if it's consistent with only one girl lying. If B lies, then B is last. Then C is lying; so *B tells the truth.* If C lies, then C is not last. Then, A, B, D tell the truth, and nobody is last! That cannot be, so *C tells the truth.* If D lies, then D is first or last. But then A or C also lied, which is no good. Thus, *D tells the truth.* The liar was A. The order of finish was $BDAC$ or $BADC$. In either case, the winner was $\boxed{\text{Barb}}$.

Problem 5-4

In a quick sketch, the inscribed and circumscribed circles are given respective radius-lengths r and R. By the Pythagorean Theorem, $R^2 - r^2 = 1$. Mutiplying by π, $\pi R^2 - \pi r^2 = \boxed{\pi}$.

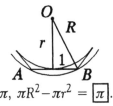

Problem 5-5

The rectangle's width is 6. Its length is 8. Call each path's width w. Since the rectangle's area is $6 \times 8 = 48$, the combined area of the two paths is half that,

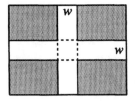

24. One path's length is 6 and its area is $6w$; the other's length is 8 and its area is $8w$. The paths overlap in a square of area w^2. Thus, $6w + 8w - w^2 = 24 \Leftrightarrow w^2 - 14w + 24 = (w-2)(w-12) = 0$, so $w = \boxed{2}$.

Problem 5-6

Recall that $-1 \le \sin x \le 1$ and $-1 \le \cos y \le 1$.

Method I: If $a = \sin x$ and $b = \cos y$, then $k = a+b-ab = a+b(1-a) \le a+(1-a) = 1$, so $k \le 1$. Furthermore, $a+b(1-a) \ge a-(1-a) = 2a-1 \ge -3$, so $k \ge -3$. Any suitable value of k must satisfy $\boxed{-3 \le k \le 1}$.

[**NOTE:** If both $\sin x$ and $\cos y = -1$, then $k = -3$; if both $\sin x$ and $\cos y = 1$, then $k = 1$. Every value of k between -3 and 1 will arise as $\sin x$ and $\cos y$ take on their other values between -1 and 1.]

Method II: As above, $-1 \le a = \sin x \le 1$ and $-1 \le b = \cos y \le 1$. Since $a+b-ab = 1-(1-a)(1-b)$, and since $a \le 1$ and $b \le 1$, we know that $(1-a)(1-b) \ge 0$. Therefore, $1-(1-a)(1-b) \le 1$. Since $(1-a)(1-b) \le (1+1)(1+1) = 4$, we conclude that $1-(1-a)(1-b) \ge 1-4 = -3$. Finally, $1-(1-a)(1-b)$ takes on all values between -3 and 1.

Contests written and compiled by Steven R. Conrad & Daniel Flegler   ©**2011 by Mathematics Leagues Inc.**

Problem 6-1

The least common multiple of 1, 2, 3, 4, 5 is the product $2^2 \times 3 \times 5 = 60$. If I increase the list to include a 6 as well, the least common multiple stays the same since the factors in 6 already appear in $\boxed{60}$.

Problem 6-2

The given equation is valid for all x, including $x = 0$. If $x = 0$, $-92 = (-1)(-2)(-r)$, so $r = \boxed{46}$.

Problem 6-3

The length of any side of a triangle is less than the sum of the lengths of the other two sides. Therefore, $\log x < \log 4 + \log 503 = \log 2012$, or $x < 2012$. The greatest possible value of the integer x is $\boxed{2011}$.

[**NOTE:** The triangle inequality cited above also tells us that $\log 4 < \log 503x$ and that $\log 503 < \log 4x$, neither of which gives rise to an upper bound for x.]

Problem 6-4

Since the three different uniform numbers add up to 21 and are equally spaced apart from each other, Bo's number, the middle number, is 7. Let the three uniform numbers be $7-x$, 7, and $7+x$. Al's number is $7-x$. Bo's new number is 6. Cy's new number is $8+x$. We're told that $\frac{8+x}{6} = \frac{6}{7-x}$, so $56+7x-8x-x^2 = 36$, or $x^2+x-20 = (x+5)(x-4) = 0$, so $x = -5$ or $x = 4$. Al's number is $7-x$. Since Al had the smallest number, his number was $7-4 = \boxed{3}$.

Problem 6-5

We're told that $b^2+g^2 = 697$. To get perfect squares whose sum ends in 7, we need squares that end in 1 and 6. Trial and error gives us the solutions $(b,g) = (11,24)$, $(16,21)$. The class sizes would be $11+24 = 35$ and $16+21 = 37$. Together, the number of kids in the two classes is $35+37 = \boxed{72}$.

Problem 6-6

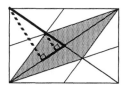

The diagonal shown splits the rectangle into two triangles, each having three medians drawn in its interior. The point at which all three medians are concurrent is 1/3 of the way from the midpoint to the vertex. Drop altitudes to the diagonal from the vertex and the point where the medians meet. By similar triangles, the altitude of each shaded triangle is 1/3 of the length of the altitude of the original triangle, so the shaded region's area = 1/3 of the area of the rectangle = $(1/3)(360) = \boxed{120}$.

[**NOTE:** Use the diagram below to show that, in the original rectangle, three adjoining unshaded regions can be rearranged to form one "unshaded" parallelogram. Each rectangle of area 360 can be rearranged into three congruent parallelograms, each of area 120.]

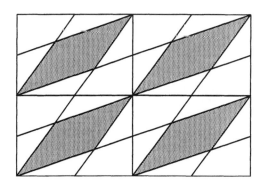

Contests written and compiled by Steven R. Conrad & Daniel Flegler   ©2011 by Mathematics Leagues Inc.

Answers & Difficulty Ratings
October, 2006 – March, 2011

Answers

2006-2007

1-1. 1
1-2. 5
1-3. A
1-4. 194
1-5. 58
1-6. $2^{11}3^7$

2-1. $-\sqrt{2006}$
2-2. 36
2-3. 6
2-4. 285
2-5. $(\sqrt[3]{3}, 3\sqrt[3]{3})$
2-6. $\pi(3-2\sqrt{2}) \approx 0.5390$

3-1. 9
3-2. 55
3-3. 4002
3-4. $\frac{486}{25}$
3-5. 18
3-6. -1

4-1. 25
4-2. 39
4-3. 74
4-4. $3^7-3^5+3^4-3^3+3^2$
4-5. 23
4-6. 1458

5-1. 3
5-2. 6
5-3. 0
5-4. 16
5-5. $2, \frac{1}{16}$
5-6. (10,90), (12,36)

6-1. 110
6-2. 142
6-3. 14
6-4. $-\frac{1}{2}$
6-5. $-2 < b < 0$
6-6. 420

2007-2008

1-1. 4015
1-2. 6
1-3. 10^6
1-4. C
1-5. $2^{2^{2^2}}$
1-6. (1,5,24), (1,6,14), (1,8,9), (2,3,10), (2,4,6)

2-1. 99
2-2. 1
2-3. 11
2-4. 17
2-5. 5
2-6. 80 or 80°

3-1. 2007
3-2. 50π
3-3. 6 or $6
3-4. 1,–1
3-5. 4951
3-6. 2, 4

4-1. 6
4-2. 9 or $9
4-3. 83
4-4. 485
4-5. (3,2008)
4-6. 120

5-1. 10 000
5-2. 13
5-3. 78
5-4. 20 or 20°
5-5. $\frac{1}{13}$
5-6. 5, 6, 7, 8

6-1. 8
6-2. 2
6-3. $\frac{1}{3}$
6-4. 4
6-5. $\frac{41}{42}$
6-6. 89

2008-2009

1-1. 2008
1-2. $\sqrt{50}$
1-3. 93
1-4. 10:47
1-5. $(2x+3y)(2x-3y+2x^2)$
1-6. 224

2-1. -3
2-2. 41
2-3. 8
2-4. 2007
2-5. -15
2-6. $\frac{1+\sqrt{5}}{2}$

3-1. 2007
3-2. 24 or 24%
3-3. 49
3-4. 72
3-5. 30 or 30°
3-6. 100

4-1. 2012
4-2. 132 049
4-3. 2
4-4. 36
4-5. $\left(\frac{2}{3}, \frac{4}{3}\right)$
4-6. $\sqrt{2}+1$

5-1. 24
5-2. 41
5-3. 47
5-4. 2
5-5. $\frac{\pi}{4}$
5-6. $\frac{1}{12}$

6-1. 4020
6-2. 26
6-3. 10
6-4. 10
6-5. $\frac{1+\sqrt{5}}{2}$
6-6. $\sqrt{24}$

Answers

2009-2010

1-1.   0

1-2.   4, 20

1-3.   75 or 75°

1-4.   (1,4,7) OR (2,1,9)

1-5.   5

1-6.   3375

2-1.   150

2-2. ± 1

2-3. $-1, 2$

2-4. $2\sqrt{2009}$

2-5.   72 or 72°

2-6.   72

3-1.   2680

3-2. $\frac{25}{49}$

3-3. $\frac{5}{11}$

3-4. $-\frac{3}{8}$

3-5.   1, 9, 10, 11

3-6.   153 846

4-1.   10

4-2. $\frac{1}{2}$

4-3.   10

4-4.   4 998 930

4-5.   60

4-6.   225

5-1.   12

5-2.   10

5-3.   half

5-4.   31

5-5.   3015

5-6.   40

6-1.   16 076

6-2.   (3,4)

6-3. A

6-4. $2x + 1$

6-5.   6

6-6. $\frac{6}{7}$

2010-2011

1-1.   2010

1-2. $(1001)^2$ or 1 002 001

1-3.   0, –1

1-4.   49 or 49¢

1-5.   126

1-6. $\sqrt{544}$ or $4\sqrt{34}$

2-1.   –50

2-2.   4020

2-3.   66

2-4.   \$8.49

2-5. $(x^2+2x+2)(x^2-2x+2)$

2-6   (19,18), (8,5), (7,3)

3-1.   0

3-2.   77

3-3.   110

3-4. $1/\sqrt{2}$

3-5. $\pi(6-4\sqrt{2})$ or 1.07802416891 . . .

3-6. $\dfrac{3}{2009}$

4-1.   1

4-2.   7

4-3.   50

4-4   8.2

4-5.   2011

4-6. $-x^3+10x$

5-1.   5

5-2.   1

5-3.   Barb

5-4. π

5-5.   2

5-6. $-3 \le k \le 1$

6-1.   60

6-2.   46

6-3.   2011

6-4.   3

6-5.   72

6-6.   120

Difficulty Ratings

(% correct of all reported scores from each participating school)

2006-2007		2007-2008		2008-2009		2009-2010		2010-2011	
1-1.	93%	1-1.	87%	1-1.	83%	1-1.	94%	1-1.	64%
1-2.	52%	1-2.	73%	1-2.	82%	1-2.	82%	1-2.	85%
1-3.	51%	1-3.	63%	1-3.	77%	1-3.	63%	1-3.	57%
1-4.	60%	1-4.	58%	1-4.	63%	1-4.	69%	1-4.	46%
1-5.	21%	1-5.	31%	1-5.	28%	1-5.	30%	1-5.	22%
1-6.	4%	1-6.	12%	1-6.	24%	1-6.	45%	1-6.	24%
2-1.	74%	2-1.	87%	2-1.	90%	2-1.	92%	2-1.	87%
2-2.	85%	2-2.	49%	2-2.	88%	2-2.	61%	2-2.	78%
2-3.	80%	2-3.	69%	2-3.	75%	2-3.	67%	2-3.	42%
2-4.	32%	2-4.	22%	2-4.	64%	2-4.	61%	2-4.	27%
2-5.	27%	2-5.	17%	2-5.	18%	2-5.	17%	2-5.	26%
2-6.	3%	2-6.	32%	2-6.	4%	2-6.	16%	2-6.	11%
3-1.	60%	3-1.	91%	3-1.	70%	3-1.	79%	3-1.	79%
3-2.	59%	3-2.	60%	3-2.	81%	3-2.	71%	3-2.	43%
3-3.	55%	3-3.	72%	3-3.	67%	3-3.	48%	3-3.	69%
3-4.	46%	3-4.	69%	3-4.	45%	3-4.	36%	3-4.	22%
3-5.	75%	3-5.	53%	3-5.	52%	3-5.	12%	3-5.	30%
3-6.	36%	3-6.	7%	3-6.	30%	3-6.	52%	3-6.	8%
4-1.	86%	4-1.	85%	4-1.	84%	4-1.	85%	4-1.	72%
4-2.	78%	4-2.	84%	4-2.	72%	4-2.	67%	4-2.	63%
4-3.	58%	4-3.	36%	4-3.	49%	4-3.	60%	4-3.	50%
4-4.	67%	4-4.	35%	4-4.	48%	4-4.	28%	4-4.	35%
4-5.	17%	4-5.	17%	4-5.	18%	4-5.	84%	4-5.	33%
4-6.	26%	4-6.	10%	4-6.	11%	4-6.	6%	4-6.	6%
5-1.	86%	5-1.	91%	5-1.	82%	5-1.	81%	5-1.	74%
5-2.	65%	5-2.	73%	5-2.	79%	5-2.	54%	5-2.	83%
5-3.	59%	5-3.	68%	5-3.	57%	5-3.	69%	5-3.	87%
5-4.	27%	5-4.	58%	5-4.	71%	5-4.	30%	5-4.	24%
5-5.	21%	5-5.	31%	5-5.	18%	5-5.	47%	5-5.	75%
5-6.	19%	5-6.	12%	5-6.	14%	5-6.	24%	5-6.	10%
6-1.	89%	6-1.	83%	6-1.	58%	6-1.	71%	6-1.	83%
6-2.	85%	6-2.	70%	6-2.	83%	6-2.	94%	6-2.	80%
6-3.	73%	6-3.	16%	6-3.	70%	6-3.	80%	6-3.	52%
6-4.	51%	6-4.	75%	6-4.	53%	6-4.	20%	6-4.	72%
6-5.	23%	6-5.	51%	6-5.	25%	6-5.	7%	6-5.	57%
6-6.	21%	6-6.	62%	6-6.	9%	6-6.	3%	6-6.	44%

Math League Contest Books
4th Grade Through High School Levels

Written by Steven R. Conrad and Daniel Flegler, recipients of President Reagan's 1985 Presidential Awards for Excellence in Mathematics Teaching, each book provides you with problems from *regional* mathematics competitions.
- *Easy-to-use format designed for 30-minute time periods*
- *Problems range from straightforward to challenging*

Order books at www.mathleague.com (or use the form below)

Name _____

Address _____

City _____ State _____ Zip _____
                                 (or Province)   (or Postal Code)

Available Titles	# of Copies	Cost
Math Contests—Grades 4, 5, 6	($12.95 per book)	
Volume 1: 1979-80 through 1985-86	_____	_____
Volume 2: 1986-87 through 1990-91	_____	_____
Volume 3: 1991-92 through 1995-96	_____	_____
Volume 4: 1996-97 through 2000-01	_____	_____
Volume 5: 2001-02 through 2005-06	_____	_____
Volume 6: 2006-07 through 2010-11	_____	_____
Math Contests—Grades 7 & 8‡	‡(Vols. 3, 4, & 5 include Algebra Course I)	
Volume 1: 1977-78 through 1981-82	_____	_____
Volume 2: 1982-83 through 1990-91	_____	_____
Volume 3: 1991-92 through 1995-96	_____	_____
Volume 4: 1996-97 through 2000-01	_____	_____
Volume 5: 2001-02 through 2005-06	_____	_____
Volume 6: 2006-07 through 2010-11	_____	_____
Math Contests—High School		
Volume 1: 1977-78 through 1981-82	_____	_____
Volume 2: 1982-83 through 1990-91	_____	_____
Volume 3: 1991-92 through 1995-96	_____	_____
Volume 4: 1996-97 through 2000-01	_____	_____
Volume 5: 2001-02 through 2005-06	_____	_____
Volume 6: 2006-07 through 2010-11	_____	_____
Shipping and Handling	$3 ($5 Canadian)	

Please allow 4-6 weeks for delivery

Total: $_____

☐ Check or Purchase Order Enclosed; *or*

☐ Visa / MasterCard / Discover # _____

☐ Expiration Date _____ Signature _____

Mail your order with payment to:
Math League Press, P.O. Box 17, Tenafly, NJ 07670-0017
or order on the Web at www.mathleague.com
Phone: (201) 568-6328 • Fax: (201) 816-0125